"Prof. Warshel's autobiography makes for compulsive reading, as it intertwines personal stories from his life with major milestones in Israeli history (as presented through the lived experience of the author) and the scientific journey that took him from his early education at an Israeli kibbutz to the Nobel Prize. This is then wrapped up with a forward-looking perspective on the direction of the field of computational biology. Overall, a strongly recommended read!"

Caroline Lynn Kamerlin FRSC
Professor of Structural Biology, Uppsala University, Sweden

"This is an exciting autobiography of one of the most creative scientists spanning the 20th and 21st centuries. Arieh Warshel revolutionized computational chemistry of biomolecules while firmly grounding the results of his research in the primacy of electrostatics as the mediator of the interactions between and within complex molecules. This book will be especially inspiring for scientists just beginning their own independent careers in its recounting of Warshel's struggles to introduce new ideas into a field not known for intellectual flexibility."

R Dean Astumian
Professor of Physics, University of Maine, USA

"Arieh Warshel gives an engaging account of growing up in a kibbutz in a country repeatedly torn by war, of the influences that led to his use of computers to elucidate complex biophysical and biochemical systems, and of the long but ultimately successful struggle to convince more established scientists of the validity and power of his approach. His memoir is written with refreshing candor and clarity."

William W Parson
Professor of Biochemistry, University of Washington, USA

"A narrative of a creative genius whose journey from a kibbutz to Stockholm is as compelling as it is inspirational. It is a must-read for young and aspiring scientists who seek to create their own path rather than follow the road well-trodden."

Devarajan Thirumalai FRSC
Collie-Welch Regents Chair in Chemistry, University of Texas at Austin, USA

From Kibbutz Fishponds to The Nobel Prize

Taking Molecular Functions into Cyberspace

Other Titles by the Author

Computer Modeling of Chemical Reactions in Enzymes and Solutions

From Kibbutz Fishponds to The Nobel Prize

Taking Molecular Functions into Cyberspace

Arieh Warshel

University of Southern California, USA

World Scientific

NEW JERSEY · LONDON · SINGAPORE · BEIJING · SHANGHAI · HONG KONG · TAIPEI · CHENNAI · TOKYO

Published by

World Scientific Publishing Co. Pte. Ltd.

5 Toh Tuck Link, Singapore 596224

USA office: 27 Warren Street, Suite 401-402, Hackensack, NJ 07601

UK office: 57 Shelton Street, Covent Garden, London WC2H 9HE

Library of Congress Cataloging-in-Publication Data

Names: Warshel, Arieh, author.

Title: From kibbutz fishponds to the nobel prize: taking molecular functions into cyberspace / Arieh Warshel, University of Southern California, USA.

Description: 1st. | New Jersey : World Scientific, [2022] | Includes bibliographical references and index.

Identifiers: LCCN 2021029996 | ISBN 9789811241789 (hardcover) | ISBN 9789811243158 (paperback) | ISBN 9789811241796 (ebook) | ISBN 9789811241802 (ebook other)

Subjects: LCSH: Warshel, Arieh. | Biophysics--Biography. | Biophysicists--Israel--Biography. | Molecular biology. | Nobel Prize winners--Biography.

Classification: LCC QH505 .W37 2022 | DDC 571.4092 [B]--dc23

LC record available at https://lccn.loc.gov/2021029996

British Library Cataloguing-in-Publication Data

A catalogue record for this book is available from the British Library.

For any available supplementary material, please visit
https://www.worldscientific.com/worldscibooks/10.1142/12412#t=suppl

Desk Editor: Shaun Tan Yi Jie

Typeset by Stallion Press
Email: enquiries@stallionpress.com

To my wife Tami for her unwavering support

Prologue

It was about 8 pm Stockholm time on December 10, 2013 and I was sitting on the stage in City Hall, with 12 other future Laureates. We were facing the King and Queen of Sweden and the Nobel Committee. After a member of the committee read my citation I stepped to the well-marked central circle, where the king handed me the Nobel certificate. I then bowed to the audience and got a very loud standing ovation. Later, I was told that this was the loudest ovation. Even the king smiled. This was an unbelievably exhilarating event for me and I kept asking myself, what was I doing in Stockholm, being awarded the top scientific prize that has been awarded in the past to the most legendary scientists? I had a hard time falling asleep that night as I tried to recapture my voyage from Kibbutz Sde Nahum to Stockholm.

Contents

Prologue		ix
I	Kibbutz Sde Nahum	1
II	The Army: 1958–1962	15
III	The Technion: 1962–1966	22
IV	The Weizmann Institute: 1966–1969	28
V	Postdoctoral Period at Harvard: 1970–1972	41
VI	Back at The Weizmann Institute	46
VII	MRC, the Cathedral of Molecular Biology: 1974–1976	51
VIII	USC: 1976–1981	61
IX	Sabbatical at the Weizmann Institute: 1981–1982	67
X	At USC: 1982–1990	69
XI	Moving Forward: 1990–2009	80
XII	Multiscale Modeling of Large Biological Systems	103
XIII	The Nobel Prize	110
XIV	Life After the Nobel Prize	115
XV	Epilogue	120
References		129
Acknowledgments		139
About the Author		141

Kibbutz Sde Nahum

The Start

The 4th day of January 1937 was an extremely rainy day and a small convoy of about 100 people, tractors and cars moved in the heavy mud from Kibbutz Ein Harod to the east. The convoy was carrying a wooden tower, barbed wire and other construction material. The rain did not let up but eventually after about ten kilometers the convoy stopped and started to erect the tower and surround it with a wall. This event was the "aliyah" (ascendance) of Kibbutz Hasadeh (the field), later to become Sde Nahum. It was the second kibbutz in the new settlement strategy called "Wall and Tower", which emerged after the Arab revolt in the British Mandate of Palestine (1936–1939). My parents were in the convoy and among the founders of the kibbutz, where I was born three years later.

My Parents

My mother, Rachel, was born in 1910 in Volodymyr-Volynskyi in what is now Ukraine. The city's Jews, however, called the city Ludmir. Her family was small and included two kids, my mother and her brother Yitzhak (Figure 1). Their father died when my mother was two years old. Nevertheless, they lived in reasonable conditions, where my mother was able to attend school and her brother went to a cadet school, becoming an officer in the Polish army.

My grandmother rented rooms in their house to teachers and one of those teachers prepared my mother for the government high school

Figure 1.   My mother (right) and uncle with their parents in Volodymyr-Volynskyi around 1914.

(Gymnasia) which was hard to get in, but it was the only free school. However, she was kicked out after two years because they found out she didn't sing the nationalist songs in the school assemblies led by the preacher. Thus she went to a private school. Subsequently she joined a youth movement that prepared her for coming to Israel. In the mid-30s she went to Israel as a pioneer. Her brother Itzhak stayed in Poland and fought the Germans upon the outbreak of World War II. He was captured and fortunately put in a stalag, a prisoner-of-war camp, with British pilots. Thus, he was not touched by the Gestapo. The rest of their family was exterminated during the war.

Figure 2. The family on my father's side (his parents, brother and sister) in Lachovitch (Lyakhavichy) around 1930.

My father, Tzvi (Figure 2), was born in 1910 in Lachovitch, which is now in Belarus. He did not finish elementary school and worked with his father, but eventually became a great autodidact. He left for Israel in 1927 to join a group that intended to build a kibbutz. My uncle, Issachar, also left Poland for Israel, where he worked in construction. He joined the British army during the war. My Aunt Chana immigrated to Israel after her brothers and joined one of the first classes of the prestigious Technion

(Israel Institute of Technology), but unfortunately decided to return to Poland and disappeared with the rest of the family during the Holocaust.

My parents met at "Kibbutzim Hill", near Nes Ziona, where new immigrants learned how to build new kibbutzim. In their case it was to be Kibbutz Hasadeh. My mother (Figure 3) worked as a beekeeper, in the

Figure 3.   My mother in the kibbutz (1940).

Figure 4.   My father (third from left) in the fishpond (1940).

production of sour cream and butter, as well as in other jobs. My father worked in construction before the foundation of the kibbutz but then became one of the pioneers in the development of fishponds (Figure 4). Years later he became secretary of the kibbutz and a bookkeeper. He was self-taught and never stopped learning.

The Children of the Dream?

I was born on the 20th of November, 1940. It was during the darkest days of World War II, but I did not know much about it. I grew up with a small group of kids my age in the Children's House, from where our parents were not allowed to take us home at night. This unique setup reflected the paradigm that the communal society should allow the workers to focus on building the kibbutz and the country. Many years later someone suggested to me that the kibbutz system collapsed in the 1970s when young mothers insisted on taking their kids home from the Children's House.

Figure 5.   Arieh in 1943 at age three.

My memories from the Children's House start around age three (Figure 5), when I was sick and put in isolation in another building. I recall seeing a giant tarantula (or perhaps it grew huge in my memory) and hiding under the blanket. I also remember my mother eventually coming to comfort me.

On weekends I frequently stayed in my parents' house and enjoyed going with my father to the Kantara fishponds. We walked about one kilometer and climbed into a small boat and distributed food for the fish. Other times we spread out copper sulfate or other chemicals to fight the different fish-killing pathogens, particularly the deadly *perimysium*.

From age three to five I had two "stepsisters" who played with me when I went to my parents' house. Rachel and Hana were among the so-called Teheran children, who were Polish kids, mainly orphans who escaped the Nazis, who found shelter in Russia and then evacuated to Teheran before reaching Israel in 1943. In our case, they were "adopted" by my parents. Eventually, however, I only had brothers.

From my time in the Children's House I remember very fondly the group of kids with whom I shared a room and the larger group of about

25 who constituted our class. I remember periods of isolation during times of seasonal diseases, and warmly recall, in particular, when my father would come to put me to sleep and read a weekly story from the national youth newspaper (*Mishmar Leyeladim*), and other kids joined me in listening. We especially liked the serial story, which later became a series of books and movies, about a special group of kids, "Hasamba", who spied on the British "occupation" soldiers.

Another memory involves the nightly howling of coyotes, which sounded very close and scary. Nights when I went from my parents' house to the Children's House I had to pass dark eucalyptus trees that made me speed up my pace.

Saturday mornings we used to wake up early to take a long walk out of the kibbutz in the direction of different wadis (valleys), named by their distance from the kibbutz, where the farthest (just before Kibbutz Beit Hashita) was called the "seventh wadi". This valley had white anemones and not just the red ones unique to the other valleys.

On other weekends we went to the old Damascus-Haifa broken train rails. The distance from the kibbutz to this train track was one kilometer and we competed in running toward the rail. Most times I would trail at the beginning of the race and then end up leading. This served me as a model in my subsequent endeavors.

For the first three years of elementary school I had difficulties in spelling, which mysteriously disappeared in the fourth grade (although it reappeared at a much later stage when I started to write in English). In those first three grades, we were divided into "good" and "bad" students and given "prizes" (colored stickers) according to our performance. I was considered the best student among the bad students, which was a somewhat problematic distinction.

Our classes included going to the fields around the kibbutz and identifying different flowers. I was good at memorizing the various names and I recall one of my classmates (Shmulik) saying that "Arieh is very good and sometimes has bright ideas, but Avital (another kid) is really bright."

Once in a while we had invasions from the north by Bedouins with their herds running over our fields. The kibbutz members were called for action, by bangs on empty gas cylinders, and chased away the Bedouins using special clubs, since guns were banned by the British government ruling back then in pre-state Israel.

I was seven years old in 1948 during Israel's War of Independence. Fortunately, we did not experience too much of the horrors of the war. We heard about the conquest of the nearby Arab city Beit Shean, almost without a fight, by using only big pipes that looked like guns. A significant number of "Youth Group III", who had joined the kibbutz as a youth movement, were recruited to the Palmach, the paramilitary organization that more or less saved Israel at the beginning of the war. Many of them died during the battles on the road to Jerusalem. Luckily the only direct damage to the kibbutz during the war was from shelling from Gilboa Mountain that blew up a seesaw in the children's playground that was fortunately empty at the time.

When sixth grade started, we moved to another building and joined a group of children a year younger than our class. We unified our groups, Eyal and Shalhevet, and called the new one Eshel. We combined most of our activities, although we had different study materials. We now had rooms for groups of four children, and the chapter of the Children's House was over.

Years later I learned that the famous psychologist Bruno Bettelheim wrote in his book, *The Children of the Dream* [1], that children in the kibbutzim had no ambition, in part because they lived together. Nothing could be further from the truth: actually, kibbutz children were very competitive, the reason many became army officers and pilots. They most probably were seeking a special identity within a large group.

Also contributing to this was the fact that the kibbutz did not encourage us to be below average or second-rate. For example, at age five we had a music expert assigned to assess our musical talent. She gave each of us a steel triangle and asked us to create a sound. After my turn, she simply told me to go away, meaning I was judged as having no musical future. She was right, at least partially. I had no musical ear and would eventually have major trouble singing from notes. Years later, however, I did participate in several Israeli choruses in Boston and Los Angeles, reaching the perfect scale by singing from memory. I even got an A in music class at the Technion. It turns out that the music screening in the kibbutz was not an effective way of instilling self-motivation. On the other hand, the American way of giving every kid a medal and always telling everyone how great they are is also not an optimal approach.

Another related point was that the kibbutz youth back then saw themselves as the most important people in Israel, and arguably the world. They would reject children from outside the kibbutz and were very snobbish to others. This haughty attitude became less prevalent in the 1970s.

Growing up, I liked competitive games, such as "Capture the Flag" (where you have to cross enemy lines and reach their flag) and related games of individual maneuvers. I also liked football (soccer), and solved the dilemma of choosing between playing and studying by finishing my homework as soon as we got it so I could go out to the field to play. Of course, many other kids never did their homework at all or waited until the last minute.

Our education did not emphasize mathematics and physics, and even other subjects did not get too much attention from most of my classmates. The real problem was that they didn't take homework and exams seriously. In fact, the main way to try to force pupils to take tests was to punish those who did not put any effort into studying by preventing them from seeing the weekly movie shown outside under the night sky. Interestingly, many years later when I heard that places like UC Santa Cruz decided not to have grades for their exams, I thought that this was a bad idea.

Our Bar Mitzvahs were held for the whole class on the same day. The day before, we were sent to perform various assignments in communities of new immigrants. At the end of the main event, we all received the two heavy volumes of the "Palmach Book" [2], about the history of the Palmach units which saved Israel in the War of Independence and then was dismantled by the prime minister, David Ben-Gurion, who was worried about the powerful left-leaning organization and its kibbutz-heavy influence, and of the kibbutz movement more generally.

At some stage, I began exploring the technical world. I started building hot air balloons and parachutes and even dropped cats with a parachute from Building A, the first two-story building in the kibbutz. I also built balsa-wood airplanes, which I tried to propel by sticking flies on the wings. I even dug a one-meter hole behind our building hoping to find treasures, which did not work too well. At some stage I began building telescopes with various lenses and mirrors. Furthermore, I built a primitive teleprinter.

I also read books non-stop on different subjects (a habit now replaced by watching television indiscriminately). I read very fast, being helped along by not trying to remember names in their exact pronunciation but in the way they sounded to me. This practice, however, didn't help my spelling of names later on in life.

Interestingly, I was the youngest boy in the class (three girls, Ilana, Gita, and Ziva were my age). This meant that I mainly felt comfortable playing with kids from the younger class. Perhaps because of this, I did not feel comfortable associating with older and more senior people when I grew up. This was probably a handicap in my ability to form political connections later in life.

I spent a lot of time with my two best friends, Dadik (David) Even-Shoshan and Uri Askebitz, from the younger class. Dadik' uncle was Abraham Even-Shoshan, editor of the most important Hebrew dictionary; and his father, Shlomo Even-Shoshan, made important contributions in translating into Hebrew Russian books (e.g., Panfilov people) that were key reading material in the kibbutz movement and the Palmach. The three of us were roommates for an extended period of time. Together we rode bikes to surrounding places and tried to sneak into the kibbutz factory for preserved fruits and sardines, trying to get some of the delicious grapefruit slices, and into the machine repair store, where we tried to use a welding torch to construct handguns. We also once stole a hunting rifle from the fishponds' storage unit to hunt ducks.

The start of high school saw no improvement in our appreciation of academic studies. I remember that the class above us used to laugh at one of the kids who was interested in science, and called him "Sinus" (after the corresponding mathematical function). Similarly, I sometimes got called the "Vilna Gaon", as the brilliant 18th-century Rabbi Elijah ben Solomon Zalman was known, as a joke about my responses in class.

When school was on, we had to work two hours a day. I started working landscaping jobs, but did not like it too much, partiuclary considering the summer heat in the Beit Shean Valley. Later I worked with the kibbutz electrician, who was originally from Hungary. I started by holding his ladder and learning Hungarian curses, moving on to more complex tasks such as passing electric wires through walls and repairing devices with heating elements.

On special weekends the kibbutz held "Recruitment" (*giyus*), where a large number of youths were "volunteered" to do "emergency" work, like finishing picking a field on a single Saturday. It was staged as a competition, and they would monitor the amount collected. I discovered I was very competitive in these picking events and vying to be the winner over what was usually one or two other competitors. This made the task much more tolerable.

I did not have a clear idea about what I wanted to do when I grew up. I remember, for example, that in one of our youth movement camps, we discussed the crucial importance of continuing to build the kibbutz. At some point I stood up and said that while I strongly believed in the kibbutz, I could not be sure now what I would be doing after the army — perhaps a reflection of the subliminal effect of my mother's hints that I might ultimately want to go to university. This might have been the reason why she insisted on making me read stories in English in the afternoons when I came to my parents' house (Figure 6).

In my senior year of high school, six of us from Sde Nahum were sent to study in the neighboring kibbutz, Ein Harod, which was founded in 1921 and was the flagship of the Yishuv Mandatory Palestine kibbutz movement. We were joining what was called the "unified class", composed of students from Ein Harod, Kibbutz Holata, Kibbutz Hagoshrim and us. This last year was one of the happiest in my life. In addition to studying, we had other end-of-school activities, including an acrobatic performance and a play. Because I was participating in both, I chose to sleep in Ein Harod rather than return every evening to my kibbutz, an arrangement I found very satisfying.

We had intensive classes in economic theory, analyzing competing socialist and communist ideas ranging from the works of Rosa Luxemburg to Comrade Joseph Stalin's History of the All-Union Communist Party. Our literature teacher, who was particularly charismatic, was Moshe

Figure 6.   My family in 1956. Upper row: my mother and father. Lower row from right to left: Arieh, Beni, Abraham and Yigal.

Tabenkin, the son of our movement leader Yitzhak Tabenkin. Unfortunately, none of this was included in the requirement for matriculation by the Israeli Department of Education. We also had physics, mathematics, biology and chemistry, but I mostly remember the less scientific classes that were taught by excellent teachers. Still, I should mention that in 2019, I read an Israeli newspaper article about the oldest active teacher in Israel, 90-year-old Zir Luz from Ein Harod, who was quoted as saying he'd had the privilege of being my physics teacher and so had contributed in some respect to my Nobel Prize.

The school year passed rapidly and the graduation ceremony was a memorable extravaganza. I still look very fondly at the photos from that period (Figure 7). Interestingly, each of us received *City of the Dove* by the great poet Natan Alterman [3], inscribed with the inspiring quote from the book, "We hope to see you carrying your nation on your shoulder."

During the summer of 1958, before joining the army, I worked in the fishponds, emptying them, transporting the carp to a small area and then lifting them into a big fish tank. We started early in the morning but

Figure 7. The unified class (Ein Harod, 1958). Arieh: first row, second from left.

lasted till late in the day, working in the mud in the very hot sun of the Beit Shean Valley. Overall, it was fine, but I never got used to the 4 am wake-up time.

At this point, despite our 12 years of studies, we had no simple path for the matriculations needed for university, since the kibbutz school was not certified for this. This meant that we had to first pass a series of "early" exams, and only then were we allowed to take external matriculation. This created a major problem and put kibbutz kids at a clear disadvantage. In fact, it is not clear to me even today what the motivation behind this regulation was. I cannot remember now what I did to address the problem of the early exams, but I recall ordering special preparatory materials that required a major amount of work and starting to do some of it in my free time.

Overall, I think my childhood on the kibbutz was very positive and happy, despite the problems associated with sleeping in the Children's House. In recent years, the kibbutz has changed completely, transforming into a system where everything has been privatized and all are left on their own. And yet the kibbutz experience has been a remarkable social experiment, which presented the purest form of voluntary common living. As in many idealistic experiments (except the Church), the kibbutz in its purest form did not last more than three generations, where the first generation was completely idealistic, the second (mine) followed, and the third stopped believing in the ideal. Still, the kibbutz movement was absolutely crucial to the foundation of Israel and its survival in the War of Independence.

II

The Army: 1958–1962

On the morning of August 12, 1958, I took a bus to Tiberias near the sea of Galilee and joined my cohorts for the "August round" of the draft, a relatively higher level than for the average recruits. We were driven to the Tel Hashomer camp near Tel Aviv and assigned to different army branches. Other kibbutz children and I were surprisingly assigned to the Signal Corps, instead of, for instance, the more elite paratroopers. The reason appeared to be that after the 1956 Sinai War it was decided that the communication specialists were not up to par and this problem had to be addressed.

After the initial assignment we were sent to boot camp, where we joined other new recruits including some who were assigned to pilot school. One of them, Uri Shani, who fell in the 1973 Yom Kippur War, was my classmate in Ein Harod, where we both had acrobatic training, and another friend, Yiftach Spector, became a legendary pilot. The three of us performed an improvised acrobatic show together, but otherwise the two-month training was not very memorable.

After this first stage, I moved to a neighboring base to start a communication course, where we trained in assorted operations, including learning to operate WWII MK101 long-wave radios. We became skilled at Morse code, training repetitively to transmit and receive faster and faster, as well as in secret codes. There were also the mundane tasks, including night guard patrols around the camp fence. We'd pass the empty pigeons' cages on these shifts, the last remnant of the Pigeons Communication Unit, which was part of the Signal Corps since before the

War of Independence and only terminated shortly before I arrived (a great tale told by Meir Shalev in *A Pigeon and a Boy*).

At the end of the course, several of my friends and I, all kibbutzniks, were attached to the Golani infantry brigade and sent to its headquarters at the Ben Ami base, near Nahariya in northern Israel. We lived in tents and worked in shifts, receiving and sending messages in Morse code, and participating in brigade maneuvers. I was sometimes assigned as the radio man for the brigade commander Colonel, Elad Peled, who later rose to the rank of general.

We also spent time on different missions. I was assigned at one point to the Syrian front, to be a temporary radio man at the 51 regiment at the Dardara outpost near what was once Lake Hula (which had been drained in the 1950s in a somewhat misguided national effort to clean up the malaria-infested swampland, only in recent years to be partially refilled). I became fascinated with the culture of the regiment, with the soldiers' special language and rough songs, even as we considered them to be somewhat primitive (we called them Mau Mau, after the violent resistance movement in Kenya). In fact, they were very dedicated and brave soldiers in a regiment that years later became a prestigious sought-after unit.

Coinciding with my move to the Golani base was my renewed interest in going to university one day. I started to carry around physics books, trying to prepare myself for future exams, usually storing them in my army kitbag. My Sears and Zemansky textbook [4] was once partially eaten by mice, perhaps drawn to the cookies that I would bring from the kibbutz. Nevertheless, I continued with the practice of taking science books with me throughout my army and reserved service including during wartime.

Not everything went so smoothly during my service with Golani. Once, during a big brigade exercise, I was manning a special communication track, which was the central station of our network, when I received a radio call from the sergeant who was our direct commander. I told him that as "the central station" I would speak first, according to the rules of radio priority. This didn't go over well with him, and he continued to insist to speak first, until I discovered my bad temper and said, "Go to hell. Over" (actually, I spat out something much worse in Russian). Subsequently I was called to our captain, Tommy, sentenced to 10 days in jail and sent to the infamous high-security Prison 4. Interestingly, I met several characters there, including the famous soldier, "Kushi" Rimon, who managed to sneak into Petra, Jordan, from where very few Israelis came back alive at that time, by stealing a UN van and pretending to be a UN soldier. My prison term ended with my realization that it is not a good idea to break radio discipline even when you have right on your side.

At some stage a few of my friends and I were summoned to a screening test for officers' course, where we took psychometric exams and had to solve problems on the spot, like crossing over a wall with a wounded soldier. We all passed.

In preparation for the course, I completely shaved my head, not an easy decision for me since I was known by some of my comrades as "the late James Dean". But it was part initiation for the course, part a way to slow down my expected baldness. My father was bald from a very young age and I mistakenly assumed that the baldness genes passed down from the father's side. (I am happy to say I am not bald today.)

The officers' course was challenging and instructive (Figure 8). It began with three and a half months of infantry training that focused on

Figure 8.   Officers' course (1959), Arieh at the front.

leadership in battle, which was capped off with being awarded the Officer's Pin. There were long navigation trips, war games, theoretical studies of past wars and, of course, an extremely arduous training regime. We learned how to conduct attacks in a platoon formation and to solve different challenges on the battlefield, where the optimal solution was the so-called "School Solution", which I figured out relatively quickly. But that almost finished off my career as an officer. Very close to the end of the course, we had a training session near an old fortress (Antipatris), and the commander of our company, the very impressive Captain Dahan, told us to prepare a plan for attacking an enemy band on a neighboring hill. I quickly figured out the best attack plan and then sat back on a rock, while others were still struggling. Captain Dahan approached me and asked loudly, "Arieh, why are you not preparing anything?" Letting my temper get the better of me, I replied, "Go and find other butterflies doing that," implying he should look for some other suckers, since I had already figured it out. The reaction to my misguided response came swiftly; Captain Dahan told me to walk back to the camp by myself. This spelled disaster and a likely dismissal from the course, one week before the end. I was extremely concerned and traveled on my last weekend vacation to the kibbutz, where I talked it over with my parents.

I am not sure I ever knew what actually happened; perhaps my parents spoke with the head of the Security Committee of the kibbutz movement, Shoshana Spector. At any rate, when I returned to the base, I only had to face a disciplinary inquiry and I was allowed to finish the course, which ended with the Chief of Staff reviewing the troops and awarding us the Officer's Pin as part of an impressive graduation parade.

We continued with about six months of communication officers' course, where we soaked up practical and theoretical information, including important fundamentals of electronic theory, and where I forged lifelong friendships.

I was assigned to a special unit that monitored the Israel Defense Forces (IDF)'s communication security — a little ironic for someone who was sent to jail for breaking radio security. Our tasks involved parking our large truck loaded with sophisticated listening and recording devices on isolated hills above major army maneuvers. One time we found ourselves the target of an attacking Golani unit. Fortunately, they weren't using live ammunition. The attacking soldiers that were completely exhausted from running up to our hill while conquering it once reached our unit, saw our radio truck, tents and antennas, and started screaming, "Food, water, women!" Zivia, the one girl in my unit, merited special mention.

Between our field assignments we stayed on the base, mostly idle, except for very long meetings with the commander and the unit staff. To pass the time, I kept up my acrobatic training, traveling to Tel Aviv for weekly classes with a trainer who used to work with the Israeli national gymnastics team. I also continued to study physics and mathematics from the books I was still lugging around.

Eventually my regular army service ended, and I returned to the kibbutz, where I had to volunteer for what was called "Third Year Service", which meant helping out on a young kibbutz for one year. I was sent to Kibbutz Yad Hanna on what was then the border with the Jordanian West Bank, and was more or less maintained by young people from neighboring Kibbutz Givat Haim. I was put to work on different field projects, and also served as "translator" for the weekly movies, where my job was to roll the Hebrew translation tape at a speed that was supposed to match the English voices. Once in a while the audience would start screaming when the translation didn't make much sense.

University was still on my mind and I realized it was not going to be simple. First, I had not matriculated but, more importantly, the kibbutz at that time demanded that for each year at the university I would have to give back two years on the kibbutz. This eight-year commitment didn't seem to be a realistic option for me given my unknown future.

Fortunately, when I was discharged from military service, Major Hillel Blaumfeld, commander of my communication officers course, told me that if I wanted to return to the army I should contact him. With the kibbutz's unfortunate requirement on my mind, I made an appointment with my ex-commander and asked to join the standing army for one year. And he accommodated my request to serve in a place where I would be able to study on my free time for the matriculation exam. What was left was the unpleasant task of telling the kibbutz secretary what my plans were, and that I would be leaving the kibbutz. He was very disappointed and told me that I would have to explain my reasons to the kibbutz General Assembly (the *asefa*) of all the kibbutz members, who would consider if my decision was reasonable. Since I was determined to reenlist, I did not see the wisdom in facing that whole assembly *and I didn't show up.* Sadly, back then the kibbutz considered my act desertion — and, in fact, for a very long time I visited my parents only via the back entrance of the kibbutz fence. Interestingly, years later most of the kibbutz members who had condemned me for my "desertion" had children *who had "deserted" Israel and lived abroad,* some here in Los Angeles probably near me.

Two weeks later I joined the Standing Army at the Chief of Staff bunker in the *kirya,* the army headquarters in Tel Aviv. My quarters were nice, at dormitories for officers and female soldiers and officers near the Yarkon River, and as I primarily had night shifts in an IDF communication center I got to spend my days studying the different subjects needed for matriculation exam. My work had interesting moments: I occasionally got to listen to the communication between the high command and the commanders on the Syrian front during the recurring ritual of border skirmishes. Once I even heard Air Force Commander Ezer Weizman responding to someone complaining that the Golan Heights is very steep: "Yes, but it is not steep for airplanes." I also helped to resolve major crises that broke out when IDF communication channels were down. I had a few minor disciplinary problems as well, particularly when Victor, the frightening Sergeant-at-Arms of the Chief of Staff complex would salute and then chastise me, saying, "Lieutenant, you can't walk here with sandals." However, most of my time passed without major incident. And I got used to city life and started to enjoy it.

Every few months I would take a matriculation exam in a different subject and I eventually passed all the subjects, although with low grades. Interestingly, during my subsequent university studies, I'd ask my friends how I could be getting such higher grades than they were when they scored so much higher than I did on the matriculation exam. It was then I learned the secret: the teachers in the city schools would tell the students the subjects that would be on the test. I, on the other hand, had to learn everything — the entire bible for the bible part of the exam, which included picayune questions about who said some obscure sentence to whom.

With my mediocre matriculation grades it became clear that I would not be accepted to Israel's top universities, Hebrew University in Jerusalem, or Tel Aviv University, which used the matriculation grade in their screening for acceptance. I had one alternative solution, namely taking the Technion's dreaded Concourse exam, with its hours of testing in physics and mathematics. I met the challenge, and was delighted to find out a few months later that I had passed.

In due time I traveled to Haifa to the Technion to register and choose a major, without any clear idea of what exactly I wanted to study. It so happened that I met Eliezer Finkman in the Technion quad. He was a good friend from officer's course who had already taken one year of physics at the Technion, which was facilitated because his more "right-leaning" kibbutz allowed its children to take the matriculation before their induction into the army. In reply to my question about what to major in, he said, "Well, you have good vision and you see colors so you should take chemistry," basing this on his assumption that in chemistry you mainly look at colored test tubes. (He had glasses since he was a kid, and thus bad vision.) With no further information, I made a fateful decision and registered as a chemistry major.

In the few months left before the start of the academic year and the end of my time in the army, I used my connections and found work at the post office center in Jaffa. My job sorting numbers and data from different forms was extremely boring, and I waited eagerly for the passing of every hour in the working day. Mundane work was the last thing I would like to do in my life, I concluded.

III

The Technion: 1962–1966

At the beginning of the 1962 academic year I rented, with a roommate whose family name was Fahima, a basement apartment in the house of Ms. Harlovski, in the Neveh Sha'anan neighborhood near the Technion. I had a very compressed schedule and an enormous amount of homework.

We were told that a grade below 2.2 would get us expelled and we also heard about the impossibility of ever reaching the legendary 4.0. Back then we wrote down all the lecture material in our notebooks, since we didn't have the relevant textbooks in Hebrew.

Some of the professors were really superb and others were less impressive, but this was a time when the students were blamed for their learning issues, not the professors. The only thing that could happen to a bad professor was to find his car inexplicably relocated to the roof of the administration building (which sometimes happened).

In that first year we hardly had any free time, as there was an excessive amount of material to study and of course we didn't want to fail the exams. But I earned very high grades and was confident I should really aim for what was considered unachievable, a straight 4.0.

During the summer break I found very repetitive work at the Phoenicia glass factory, using a machine to print the word "Stocks" on drinking glasses. Fortunately, at the end of the summer I was awarded a modest fellowship and moved to the students' dormitories ready to start Year 2 with less anxiety.

I joined the gymnastics club at the beginning of the first semester, which I continued religiously until the end of my fourth and final year. I enjoyed dorm life and made some money by agreeing to wake up a friend who lived in another room as well as tutoring an African woman who was in Israel as a guest of the Foreign Ministry. More importantly, I excelled academically, and started to get straight A's. And I found that I enjoyed some of the lectures and the challenges of solving interesting differential geometry problems.

At some point in 1963, I was called up to the army base at Kordani near Haifa and assigned to a tank regiment as a communication officer by a Colonel. I protested that I had much more sophisticated training with Morse code and secret code, but he said that he can send me to a course of regiment communication officers, and I instantly said no thanks. He also mentioned that I would leave the tank core only on a stretcher.

In the summer of 1964, after my transfer to the armor core I was called to a multi-brigade exercise in the Negev, where I was told I was replacing a communication officer who had fought in the1948 War of Independence and had the distinction of helping in erecting the legendary "Inc flag" that was improvised with an Inc drawing and raised when the Negev brigade conquered Umm Rashrash, today's Eilat. My regiment commander was Shaul Yafeh, who had been the first commander of the Israeli armor units, but was dismissed from the standing army by Ben-Gurion, who did not trust the left-wing Palmach commanders. At any rate, our tank was a place of pilgrimage for all the top armor commanders who grew up under Yafeh.

On my 24th birthday, on November 20, 1964, my mother sent me a cake from the kibbutz. For my birthday party, held in the dorm, my classmate Hana Shanna (later Hanna Shvo) invited her cousin Tamar (Tami) Fabrikant. I met Tami again a few weeks later, on a Friday night when I went to the club at the old Technion building. After dancing all evening, we started dating and Tami and I became a couple.

The third year marked the start of my interest in biology. I was fortunate to land a research project with Professor Yechiel Shalitin, exploring enzymatic reactions. Enzymes are protein molecules that participate in virtually all bodily functions by accelerating the relevant chemical reactions. Some enzymes act as scissors that cut the bonds that connect parts

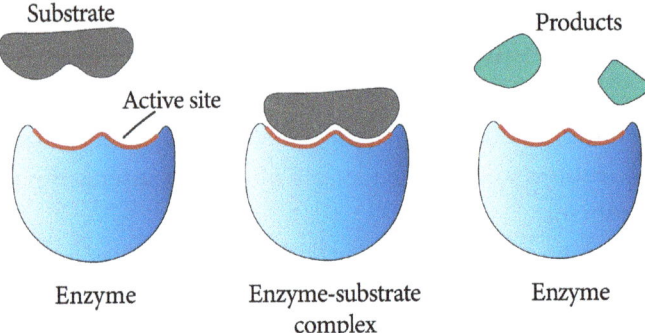

Figure 9. A schematic description of how an enzyme cleaves the molecule that serves as its substrate.

of their substrates (Figure 9). It used to be assumed that substrates fit the enzyme like keys fit their corresponding locks. However, this simplification was incorrect and could not provide any way to assess why different enzymes have different catalytic power. The great question was what made enzymes work so fast. I was assigned to study chymotrypsin, an enzyme secreted by the pancreas that catalyzes the chemical reaction of breaking bonds in proteins and peptides. My project involved using a pH meter, which measures the acidity, to monitor the rate of that reaction. This led to what turned out to be the incorrect conclusion that electrostatic effects are not very important in enzyme catalysis, since the amount of salts in the test tube did not influence the rate in a significant way. It would take me a long time to gain a deeper understanding of electrostatic effects, but at this stage I became fascinated by the prospect of measuring fast reactions. I started to study relaxation methods and other approaches that could help in evaluating the rate constant of the first step of the chymotrypsin reaction. I read about the use of line broadening of nuclear magnetic resonance (NMR), and discovered that the only NMR machine (NMR 60 MHz), which the Chemistry Department had, could not yet perform that job, but that the Computer of Average Transients (CAT) the department would soon receive might help in obtaining a decent line shape. After the arrival of the averaging device, in 1965 I obtained a line-broadening picture that allowed me to determine the rate constant. This was an accomplishment that, to the best of my knowledge, had not been reported at that time by much more established scientists, but unfortunately I declined Shalitin's suggestion to publish the results because I thought the work was not yet clean enough — not a very wise decision.

I also gradually learned about wider aspects of biology, and volunteered to give a seminar on allosteric modulation, in which biological systems are regulated by a feedback mechanism. This was an opportunity for me to learn about the work of François Jacob and Jacques Monod who, together with André Lwoff, received the Nobel Prize in Physiology that same year.

I also took a physics class taught from Merzbacher's quantum mechanics textbook [5], which was challenging for me as a chemistry major. I noted that the professor kept talking about "asymptotic approximations" like the Born approximation, whereby one solves complex problems by looking at solutions at infinite distance, instead of looking at a short distance where the solution is very complicated. Then I told a physics student that one day I was going to find an asymptotic wave function for enzymes. I didn't have any idea what I was talking about, but 15 years later, when I developed the empirical valence bond approach, I managed to do exactly what I "promised" back in 1965.

That summer, I was lucky to land a job in the Department of Earth Engineering. My bosses were two young professors, Benjamin Zur and Alexander Mukadi (who was later killed in the 1967 battle in the Golan Heights as a tank brigade reconnaissance officer). Both were pioneers in the emerging field of drip irrigation, and my assignment was to build a special pressure cell with different compositions of clay and cellulose that would form a semipermeable membrane for water flow. After measuring the effective water flow in the lab, I had to evaluate the corresponding diffusion constants. This task involved taking the results of thousands of measurements and calculating the diffusion constants. I performed the calculations with a mechanical calculator, sitting and crunching the numbers five hours a day, which somehow didn't bother me. But one day a theoretical quantum chemist who was a teaching assistant in our class told me that they were using a computer with paper tape for their calculations. It was enlightening to learn that if I could only type my data and write a short program, I would find the diffusion constant instantly. This realization occurred close to the end of my summer job, so I didn't try to look for access to that computer. However, I registered this important information in my head for future use.

At year's end, I was awarded the title of the Best Third Year Chemistry Student and got a special certificate from then-prime minister Levi Eshkol. I noted his missing finger at the time, which I later learned he lost in a work accident. Overall, Eshkol was a great prime minister who never got enough credit for his enormous contributions to the foundation of Israel.

By the middle of my fourth year I started to look for the next step in my career. One option was to continue in the M.Sc. program at the Technion. This entailed taking three language classes, including English, the only class where I didn't get an A. I started to take evening Russian classes but this language requirement and the fact that the M.Sc. was a three-year program led me to look for some alternatives.

Somewhat indifferently, I sent a few letters to American universities, asking for application material. Then one day I read a story in the weekend Yediot Aharonot newspaper about the scientific director of the Weizmann Institute of Science, Shneior Lifson, and learned he was among the founders of Kibbutz Tel Amal, now Kibbutz Nir David, which was about three kilometers from my kibbutz, Sde Nahum (Figure 10). Tel Amal was the first "Wall and Tower" kibbutz, founded at the end of 1936, and my kibbutz was the second. I learned to swim in the Asi River that passed through Tel Amal. Intriguingly to me, the article also mentioned that Shneior rode a bicycle to work. I was very impressed and decided I wanted to meet him. I made an appointment and took a bus to the Weizmann Institute in Rehovot with Tami, soon to be my wife. Shneior's secretary took us to his office, and at some point I dropped my list of

Lifson (Nir David) ----- Warshel (Sde Nahom): about 3 km distance

Figure 10.   Nir David, where Lifson was among the founders, and Sde Nahum, where Arieh grew up.

grades on his desk. Written on yellow tape was a tally that showed a near perfect 4.0. He glanced at it dismissively and told me he didn't care about grades, on top of which he wasn't accepting any students. However, he was evidently impressed and made an exception. I was accepted.

IV

The Weizmann Institute: 1966–1969

A Short Master's

On August 24, 1966, at the end of the academic year, Tami and I got married. After a short honeymoon in the northern town of Nahariya, we moved to the Weizmann Institute, initially into a mini dormitory called "Sagan House" and then to bigger student housing, the "Clore House". Named for the great benefactor Charles Clore, this building was notable for the guest who asked why there was a photograph of Vladimir Lenin on the wall, referring to the image of Israel's first president and the Institute's founder, the similarly bearded Chaim Weizmann.

I immediately joined Shneior's group of researchers, which included Mordechai Bixon, who was then finishing his PhD. Shneior had recently moved from the study of statistical mechanics of helix-coil transition and polyelectrolytes to a new field that reflected his emerging belief that using computers for treating molecules at the atomic level would offer more promise.

Bixon, who became a deep and smart scientist, started the computer modeling project, using internal coordinates (bond lengths and bond angles) for treating ring-type alkane molecules. A pioneering work of Bixon and Lifson resolved the puzzle of the observed X-ray of the ring molecule cyclodecane, which appeared to reflect two conformations, and demonstrated the potential importance of the use of force field modeling.

My proposed research would be aimed at the development of force fields that could be used for modeling proteins. Learning to program in FORTRAN, I tried to apply Bixon's program to cyclic molecules with a peptide bond, which defines the main chain in proteins (such cyclic molecules are called lactams). In fact, we already knew of an attempt to generate a protein force field by an Italian scientist, Prof. A.M. Liquori, but his work was not based on any refinement procedure and unfortunately for him he was rumored to be hospitalized in a psychiatric facility. At any rate, I started to write a general program to refine the lactams' force field parameters.

Unfortunately, trying to implement Bixon and others' formulation and use internal coordinates for cyclic molecules involved a complex dependence between the coordinates (where, for example, changing bond length could change the torsional angle). This made it very hard to obtain the correct first derivatives of the energy, which were needed for finding the most stable minimum energy structures. Trying to determine whether the first derivatives were correct, I repeatedly checked the code, sometimes finding mistakes; I had growing doubts about whether this approach would ever work. Furthermore, at some stage Shneior casually suggested that I also calculate vibrational normal modes, which would allow extracting direct information about the forces between the bonded atoms. This required the second derivatives, and I realized there was absolutely no chance of getting the relevant analytical second derivatives without errors. The formula for this was extremely complicated in cyclic molecules due to the dependence of the coordinates on each other. I remember showing the pages with the code to my youngest brother Beni when he visited me at Clore House, lamenting how complex and hopeless it was. At that stage, I tried to optimize the structures of the lactams, even without fully correct first derivatives, which led to reasonable minima, but clearly they weren't exact.

In parallel with my research I took M.Sc. classes, studying chemical physics, mathematics and biophysics. I remember Ada Yonath, who went on to win the 2009 Nobel Prize in Chemistry, teaching us X-ray crystallography and giving her a hard time with my complicated questions. In a similar vein, an exam in biophysics, a course taught by Professor Shimcha, who was on a sabbatical at Weizmann, led to my rebellion. That is, Shimcha did not know Hebrew and wanted to administer the test in English. I, however, demanded it be in Hebrew, arguing my one B in English *was* enough. Eventually we agreed on a compromise: the exam would be translated into Hebrew by another professor.

Desperate with what looked to me like slow progress in getting first derivatives, I had the idea to use the computer as my teacher (which became key to my philosophy) instead of looking for bugs in their expressions. So I turned to the "Golem", Weizmann's flagship second-generation computer, named for the legendary medieval "robot" created by the 16th-century rabbi of Prague to defend against anti-Semitism. It was designed with 18 significant digits according to the specifications of the famous, and very intimidating, mathematician Chaim Pekeris, who always got his way in getting thing done. Shneior told me confidentially, for instance, that when he was the scientific director of Weizmann, the construction manager came to him crying, complaining Pekeris had forced him to sign a letter stating that "Today (a day one year from now) he is resigning because he did not finish construction of the new structural addition to the computer center." Of course he finished on time.

I basically was "afraid" of Pekeris, but one day when I was studying in the math library, he approached me and asked if my father worked in the post office in Jerusalem. I responded that he worked in the fishponds in Sde Nahum. This was to be our only conversation, but at least he was friendly, and not frightening anymore.

Before returning to the Golem, it is interesting to note that in the early 1950s an international committee was assembled to decide whether the Weizmann Institute needed a computer. A senior member of the committee, Albert Einstein, asked, "Why did such a small country need such a big computer?" but John von Neumann argued successfully that Israel did actually need one, leading to Weizmann acquiring its first computer in 1954, the Weizak.

Working with computers in the sixties and the early seventies was completely different to what it is today. The program was on punched cards, an advance from the previous era's paper tape, and every change had to be punched in on a new card. Then the very large card box was fed to the computer, and we had to wait several hours for results. Even minor errors required resubmission, so turnaround was maybe twice a day. This meant that every evening around midnight, the end of a date with Tami, for example, I would stop at the computer for a chance at another round. Of course, this required major patience from Tami.

Now back to the Golem. Realizing it had such a high numerical precision, I decided to use Cartesian coordinates (x, y, z), where there is no complex inter-coordinate dependence. By making very small changes in the coordinates, I could get numerical derivatives and compare them to the analytical derivatives, which were easily evaluated. To my great surprise, this appeared to be an extremely powerful strategy, which provided me with an easy way to see if my analytical derivatives were correct or incorrect. In particular, I saw that when the potential is expressed in terms of Cartesian coordinates, I could suddenly get simple first derivatives that were correct — they had the same analytical and numerical derivatives — even for cyclic molecules. I also calculated numerical second derivatives from the analytical first derivatives (because of the Golem's very high precisions) and started to get the vibrations of cyclohexane from my force field. The vibrations were reasonable, but they had some strange features in terms of symmetry.

In my naivety, I tried asking some physicist colleagues if they had any clue why this was happening. However, as turned out with most of the problems I have approached since then, I couldn't get help from anyone except myself. Eventually I discovered that the problem was in assigning the origin on one of the atoms of the molecule, instead of on a point in free space.

At this stage I started to write a Cartesian force field program, and told Shneior that I had basically solved the major challenge in the field and so should begin my PhD immediately, after only half a year working on my Master's. To my delight, he agreed, despite this being unprecedented. I simply would have to write a summary of my lactams' minimization as my M.Sc. thesis as well as a PhD proposal. I completed both and moved on to my PhD.

The Six-Day War: June 1967

On May 14, 1967, in a series of aggressive moves, Egyptian President Gamal Abdel Nasser mobilized the Egyptian army in Sinai; expelled the UN force on May 18; and blocked the Tiran Straits on May 22. On May 16, there was a knock on our door at the Clore House and a driver from my military reserve unit entered and told me that we had to travel to the brigade's base in the north. This was the start of "the waiting period", when Nasser blocked entry to the Red Sea and was primed, as were Syria, Jordan and Iraq, to choke off Israel. Despite what was effectively a declaration of war, none of the nations of the world offered any useful help for breaking the blockade, leading Israelis to feel an existential danger. The mobilization of the reserves did not reduce the anxiety in Israel.

Although our brigade was assigned to the crucial Syrian front, on May 26 most of the brigade forces left for the Jordanian front, and only my 377th tank battalion remained in the north. In my role as communication officer, I was assigned to the tank of the battalion commander, Lt. Col. Amnon Hinski, who replaced the legendary Shaul Yoffe. Plans were drafted to take the source of the Banias River (the Banias heights), on the lower slopes of the Golan Heights, without any additional help, including a rather crazy plan of attack in the style of "The Guns of Navarone", climbing with ropes on the steep Hamra hill plus traversing a deep anti-tank ditch. A special crane was supposed to throw a large log into the ditch that would enable our tanks to cross over it. In trying this idea we dug a ditch and I entered the tank that started to move over a large log. At the last minute I decided that it was better if I would not be in the tank and jumped down. Then I witnessed the tank spending a very long time on the log with its bottom exposed for a sufficient amount of time to destroy it with a bazooka. Thus it seemed that traversing an anti-tank ditch while moving over a log was not such a great idea. Fortunately, we never executed this attack plan.

On June 5, the IDF code "Red Sheet" was broadcast on the fighting units' radios and the Six-Day War was launched. The destruction of the Egyptian air force led off, followed by ground battles in the Sinai desert and the Jordanian-held West Bank. At this stage, my unit had done very little, aside from clashing with a company of Syrian tanks that tried (very unwisely) to attack kibbutz Dan. The enemy tanks were destroyed and we were idle for five more days.

At the end of these five days, the Arab armies on the southern and central fronts were ready for a ceasefire, but the Northern Command was eager to attack the Syrian forces. At one point I heard the Chief of the Northern Command David Elazar (Dado), later the Chief of Staff, saying the Syrians would flee like locusts when they hear people hitting empty tin cans. Defense Minister Moshe Dayan objected to the attack on the Syrians but perhaps due to pressure from the northern kibbutzim, he relented, and called for attacking the strategic Golan Heights. And so late in the morning on Friday, June 9, our tanks started to move from the Kiryat Shmona high school toward the attack position. At that point, our 377th battalion had been put under the command of the 8th Brigade, commanded by Col. Albert Mendler, who had come from the southern front with one battalion, the 129th, to lead the attack. Movement began with the 129th leading and then our 377th, significantly depleted after "loaning" two companies to the Golani Infantry Brigade, headed by a fighting unit, composed of our "P" company, the battalion commander's tank and the tank of his lieutenant commander, Benzion Padan. Unfortunately, the 8th Brigade was new to the Golan front, and the Northern Command's leading reconnaissance officer, who was supposed to best know the Golan Heights, committed a major error when he missed the turn east and pulled us south.

The commander of 129th, Arieh Biro, moved approximately in the correct direction with the brigade reconnaissance officer, Alexander Mokadi (who was my supervisor in Technion's Soil Science Department in the summer of 1965). When Biro was wounded and evacuated backward, Mokadi tried to take the command, but he was killed when his tank was hit and turned over; his body was not found for another week.

The leading company continued to move, finding itself in a major killing field below the village Kala, where it fought heroically most of the day (leading to several citations for bravery, including the highest, the Medal of Valor), reaching Kala with only two tanks. At the same time, Col. Mendler, now aware of his initial error, tried to push the remaining troops, including our small fighting unit, toward the original attack direction of the village Zaura, a problematic attempt mid-battle to alter the paths of the tanks and the target. From my vantage point in the battalion commander's tank, I witnessed confusing commands and communication challenges of unclear attack directions. At some stage I responded to the colonel's radio call and suggested that he communicate directly with the commander of the P company, since our battalion commander was busy shooting with the tank machine gun. But we moved on gradually in the correct original direction, led by the lone tank of the lieutenant commander, who was eventually wounded when his tank was destroyed by RPGs. A few P company tanks trampled the Syrian defensive positions at Zaura. They then moved to Kala and arrived there before the 129th force that was still fighting heroically trying to move uphill. In the evening we and the 129th force joined at Kala. The Syrian line on the Golan Heights collapsed, and we continued on the next day and took the city of Kuneitra. We encountered only minor resistance and no major events, except that I got shot in my right ear and my helmet exploded. Fortunately, I ended up with only a scar.

Near Kuneitra we captured a Russian communications truck that had its own vacuum cleaner. Then Mendler, the brigade commander, had asked to see the truck after he already visited a capture T55. He entered the truck and then left. So when the commander of P company, Avraham Noifeld from Kibbutz Kfar Giladi (Figure 11), asked my permission to take home the vacuum cleaner as an excellent present for his wife, I told him it was fine with me, and Noifeld drove off in his Jeep to his kibbutz, which was not too far away. Unfortunately, I was told that Mendler wanted to check out the truck again, and in great panic I alerted Noifeld, who sped back to his kibbutz to retrieve the vacuum cleaner. I would have a hard time explaining what became of this precious Russian household appliance had Noifeld been late.

Soon after the war I was called to the brigade office of the lieutenant commander, who inquired about a rumor that I had at one point intentionally turned off the radio in the battalion commander's tank. I responded that I simply suggested to Mendler to open a direct line of communication with the P company commander, since our commander was otherwise occupied with the machine gun. Years later I realized that

Figure 11. With some of the 337th command crew after taking Kuneitra, 10 June 1967. Arieh (in the center, looking back and holding a map) talking to Abraham Noifeld, the commander of the P company, who was killed in the 1973 war as a regiment commander.

this attempt to blame our commander for the confusion that resulted from the navigation error may have been part of what became known as the Generals War. At any rate, Col. Albert Mendler received most of the credit for taking the Golan Heights and was promoted quickly. In the 1973 Yom Kippur War he was the commander of the regular army division in the Sinai front, and was killed by a direct hit on his armor half truck. A major general at that point, he was the most senior Israeli casuality of that war. As to the 1967 war, the High Command came to term that Golan navigation miscalculation "the blessed error", although it was in fact a very serious mistake that probably cost many casualties.

The PhD and the Consistent Force Field

At the end of June, we were released from the army, and I returned to the Weizmann Institute and my PhD. Incidentally, at the beginning of the war, when the air raids first sounded, Tami had thought to grab my M.Sc.

manuscript and my doctorate proposal as she ran to safety to the Clore House basement. Fortunately, all was calm as the Egyptian air forces had been swiftly destroyed.

I was ready to restart my PhD project, based on the Cartesian concept and the corresponding simple conformational program I had written, along with my earlier refinement program. I also started working on the use of group theory for automatic assignment of molecular vibrations from calculated normal modes, for the purpose of refining the force field.

In the fall we had a new addition to our group, when Mike Levitt arrived from England. Mike, who had grown up in South Africa and then moved to England, had tried to join the Medical Research Council in Cambridge at the end of his undergraduate studies, but was told by Sir John Kendrew that he should first spend a year at Shnieor's lab. He arrived at Clore House where at the start he was very well dressed, with silk shirts that he ironed himself, a fashion trend that completely disappeared in subsequent years.

I was somewhat reserved when Shneior asked me if I wanted to let Mike join my project, telling me how great more manpower would be. In particular, I did not want to share the credit for the advances I made in finding out the enormous power of writing a program in a Cartesian representation. But after a while I agreed to meet Shneior and Mike (Figure 12) in the Clore House lobby, and we talked about what might be our general approach, including the idea of a special code for molecular formulas and a Cartesian-based program, which could now be written more cleanly.

I instructed Mike about the general requirements of the program, with the analytical first and second Cartesian derivatives. He started to apply his programing talent and we were seeing progress. In parallel to checking the program, I implemented the systematic improvement of the force field parameters. This required specialized advances in obtaining the derivatives of the calculated molecular energies, molecular structures and vibrational frequencies with respect to the force field parameters. Having these derivatives allowed me to refine the force field parameters by the least-squares procedure. This led to the best agreement between the calculated and observed properties. I believe that today we would describe

Levitt

Lifson

Warshel

Figure 12.   Lifson, Warshel and Levitt at the Weizmann Institute (1967).

our approach as a kind of machine-learning strategy. Shnieor named our procedure the Consistent Force Field (CFF) method, and I began to summarize the approach and the results for a paper for publication. Considering my less-than-stellar level of English, I was grateful to accept Shneior's invitation to his home to get writing help several days a week. He and his wife Hanna were superb hosts and I became a frequent visitor, a benefit that allowed me to meet the important people who were regular guests, including members of the Labor Party and the country's cultural scene. I learned to appreciate Shneior's interest in culture, which occupied a significant part of his time, and I enjoyed his way of telling stories and jokes, even if he sometimes repeated the same stories. The CFF paper, my first, was finished and published in 1968 [6].

Subsequently, in an attempt to obtain better refinement of the parameters for the interaction between non-bonded atoms, I also developed a program that evaluated the energies, structures and vibrations of molecular crystals [6]. This required learning about different models of treating molecular motions in crystals and evaluating the corresponding vibrations (phonons). I carefully read Born and Wang's classic textbook [7], but as with other cases, I found that my Cartesian treatment made the formulation and calculations almost straightforward. I was able to calculate at the same level the inter- and intra-molecular vibrations, a task that no other approach could accomplish at that time.

The best demonstration of this occurred while working with Otto Schnepp, who was visiting Weizmann. Otto, a refugee from Vienna who grew up in Shanghai during World War II, studied for a PhD in Berkeley, then emigrated to Israel and was one of my teachers at the Technion. Subsequently he moved to USC, but came back for a sabbatical to the Weizmann Institute in 1968. A world-leading spectroscopist of molecular crystals, he was trying to refine the intermolecular forces between N_2 molecules by fitting calculated and observed vibrations in N_2 crystals. He was using the internal coordinate methods common in the field, but it was not clear how to implement those methods in examining the effect of shifting the interaction center from the position of the nitrogen atoms along the N–N bond, and so solving this problem seemed to be extremely challenging. One day I told Otto that I could do it very easily with my Cartesian program. He was skeptical, but we ended up refining the force field and writing a nice paper on the subject [8]. A few years later, Otto wrote a review on the subject of modeling crystal vibrations [9], in which he considered studies of inter- and intra-molecular motions without mentioning my approach. Instead he described an extremely complex Japanese method that used internal coordinates or approaches that used Euler angles. I do not believe Otto wanted to reduce my contribution, but he couldn't really believe that such a complex task that required such enormous effort became almost trivial with computers and Cartesian coordinates.

The development of the Cartesian program taught me an important lesson about the power of computers. I realized that normal mode calculations that required a year of training in the so-called FG method became just one formula whose solution was almost instant with our Cartesian program. Of course, this required digital computers. In other words, many problems that called for developing sophisticated methods could now be solved by considering their clear initial formulation (before the subsequent specialized formulation) and letting the computer do the work. Obviously, this approach looked strange to those who had spent great effort learning specialized methods and were told that computers made these methods obsolete.

Another interesting project, which ended up being the first quantum mechanics + molecular mechanics (QM+MM) study, was a collaboration with my friend Abraham Bromberg. His PhD project involved a study of the kinetics of an oxygen attack on dihydrophenanthrene, where he found that the observed kinetics could only be described with an enormous nuclear tunneling correction. In an attempt to justify the experimental findings, we tried to obtain a realistic potential surface for the reaction and to calculate the correction. We would have to obtain the vibrational frequencies of a large molecule and represent the reacting region by a quantum mechanical description. Our solution was to represent the reacting region by a complex valence bond quantum mechanical Hamiltonian and to describe the rest of the system by the CFF force field treatment [10]. This reproduced the experimentally deduced large tunneling, although some workers could not believe that we actually evaluated the normal modes of such a large molecule.

In 1969 Mike and Shneior published a pioneering energy minimization lysozyme [11] demonstrating the power of our computational approach. I declined Shneior's repeated suggestions to be included as an author considering the fact that I did not contribute to this important paper. Subsequently we refined CFF parameters for peptide bonds [12].

In the middle of all this, Mike asked me to be a witness in the Tel Aviv rabbinical court for his wedding to his bride Rina. A requirement for a Jewish wedding, two witnesses testify about the future husband's good standing, including that he is not already married. I happily obliged, and sometime after his wedding, Mike moved back to Cambridge to start his PhD.

In 1969 Shneior was awarded the prestigious Israel Prize, which was very well deserved. Before the ceremony, Shneior asked me to join him at his worktable during an interview with Israeli TV, which was still in its infancy. Tami and I attended the impressive event, where Shneior spoke in the name of the other winners. A short time later, I was called up for reserve service and Benjamin Bar-Lev (brother of IDF Chief of Staff Chaim Bar-Lev), who was our armaments officer, greeted me saying, "We just saw you on TV. You really looked like Robert Taylor." I took it as a great compliment.

Toward the end of 1969, I finished my PhD and looked for a post-doctoral position. I considered to work with Martin Karplus from Harvard, who had spent half a year on sabbatical at the Weizmann Institute, where he'd tried to head in a more biological direction after gaining prominence in theoretical chemistry. I'd been his host on several occasions, including joining him, his wife Susan and their two young girls, on a lovely trip to Sharm El Sheikh at the southern tip of the Sinai Peninsula. That was my first exposure to American politics, when at the ruin of the Nabatean city of Avdat, I watched Karplus waving a stick like a sword, declaring, "We're going to kill Nixon." Later on, I was also able to be very helpful to Susan with her own project on vibrational calculations. Karplus offered me a postdoc and I accepted.

V

Postdoctoral Period at Harvard: 1970–1972

I took a two-week tour of Europe, my first time abroad, while Tami, who was pregnant with our first child Merav, stayed in Israel. Subsequently I arrived in Cambridge during a snowy winter, where you could not survive outside for more than five minutes without covering your face with something extremely warm. I stayed a few days with Barry Honig, who was also a theoretical chemist in the Karplus group (Barry later became a rival but ended up a good friend). Thereafter I rented an apartment just outside Harvard Yard and started my postdoctoral period.

I joined Martin Karplus' group and was given a workspace at Prince House, a wooden building where Timothy Leary had experimented with LSD in the early sixties. I arrived long after Leary was fired from Harvard and when Prince House was merely hosting theoretical chemists of the Karplus and Roy Gordon groups.

My main project was aimed at extending the idea of combining force field with quantum mechanics, which I had already started with the dihydrophenanthren project. Now, however, the idea was to model conjugated molecules, where the π electrons (the electrons in orbitals perpendicular to the bond) would be treated quantum mechanically and the σ electrons (which hold the bonding skeleton) would be treated by a classical force field.

In parallel I advanced a project where I tried to explore scattering an alkane molecule from an alkane crystal, in the hope that having fixed molecules in the crystal would allow us to get clearer information about non-bonded potentials. The theoretical treatment involved representing the molecules by spherical harmonics functions (which describe motion of a vector in two angular dimensions) and then moving to a third dimension by using Wigner's matrices. This sophisticated theoretical treatment finally worked and I got a detailed picture of the scattering process. Unfortunately, I found out that despite the clear non-uniform nature of the interaction forces, I was getting what is called "specular reflection". This means that the angel of reflection is equal to the angle of collision (like a tennis ball hitting a wall). The idea of getting more information about intermolecular forces by scattering from crystals did not seem promising anymore. Here, as is the case with many scientific projects, I found out that the amount of work invested does not guarantee getting the expected rewarding results. However, at least I widened my theoretical background.

I enjoyed life in Cambridge outside my scientific work; my activities included weekend family trips, especially to the New Hampshire ski resorts. When Tami arrived with the six-weeks-old Merav, she had insisted strongly that we move to better housing that didn't feature corrosion in the bathtub. Fortunately, we received nice accommodations in faculty and postdoc housing in the "Botanical Garden", where we enjoyed the company of other postdocs and visitors.

This was the time of the students' protests and it was interesting to watch the demonstrations as well as concerts on the Cambridge Common (Figure 13). There was a dart board at Prince House with the face of President Nixon, which was the target of the postdocs and PhD students, who were very vocal about the invasion of Cambodia and other hot-button issues. The argument that China would get involved following the invasion was not of great concern to me. In fact, I viewed the invasion from my Israeli perspective of active response on terror operations.

Looking for inspiration, I spent a lot of time in the library and glanced through many scientific articles, exploring different possible directions. It was there that I got one of my most important ideas. This happened after reading Tully and Preston's paper [13] about their semi-classical surface-hopping approach for a system of three atoms, in which

Figure 13.   Saturday at Harvard Common in 1971.

one runs classical trajectories (by solving numerically Newton's equations of motion) on the energy surface of a given electronic state and evaluates the quantum mechanical probability of jumping to another surface. This approach enables getting the quantum mechanical time-dependent probability of jumping between electronic states (the so-called inter-system crossing) without having a quantum mechanical representation of the enormous number of vibrational states. At that time, I was very interested in photoisomerization (the rotation around double bonds of a conjugated molecule, following absorption of light), aiming at an eventual understanding of the first step in the visual process. However, all the available approaches involved quantum mechanical description of all the vibrational states and evolution of the probability of radiationless transitions between these states, which was impossible to implement in any reliable way. More importantly, as I found out later, such approaches led to the wrong results since they did not include the direction of the quantum wavepocket, which described the molecular internal motion. At any rate,

I assumed intuitively that the surface-hopping approach should work better for systems with much larger dimensions than a triatomic molecule, since in such systems the energy levels are very dense and probably follow the classical trend. I realized that if this assumption turned out to be correct, we would finally have a powerful way to study key chemical processes in condensed phases, including in proteins. Interestingly, at that time I also developed a unique approach that allowed me to evaluate all the vibronic transitions (transitions between vibrations in different electronic states) in large molecules in the hope of using such approaches in studies of photoisomerization. Fortunately, I had the semiclassical approaches as an alternative.

Thus I decided to explore the surface-hopping approach in studies of photoisomerization reactions. This was an example of moving in a direction that the hard-core chemical physicists would consider far too speculative. Of course, I spent a lot of time exploring the formal justification of the semiclassical trajectory approach, in particular studying Bill Miller and the Russian papers. My main motivation was still based on an intuition. Adapting an idea from the field of atomic collisions was an example of the enormous benefit of finding relationships among distant problems.

Another good example of this was my realization of the way of obtaining quantum mechanical analytical first derivatives of the potential energy of a molecular system. That is, during the development of the QM+MM π electron method, I strived to retain the original features of the CFF approach, including having analytical first derivatives (such derivatives are very useful for energy minimization). Yet this seems to require differentiating the quantum mechanical wave function that would have been extremely challenging. Intuitively, I tried to hold the wave function (represented by what is called the bond orders) fixed and to only differentiate the integrals that represent the Hamiltonian (represented by analytical functions). Using my old trick and comparing the numerical derivative to the one obtained by my approximation, I found to my surprise that my treatment was exact.

I later ascertained the reason for the validity of my finding, but the important point was that I discovered, by using intuition and a computer, a very important practical rule. This can be put into perspective by noting that in the mid-80s, what had basically been the identical finding by Car and Parrinello (that was "sold" by considering the wave function coefficients as masses that were treated by fancy Lagrange multipliers) was considered one of the greatest advances in theoretical chemistry. Interestingly, when I presented my finding to Martin Karplus, he first refused to believe it, but later said, "Of course, it's the Hellman-Feynman theorem." Indeed, it is true that my results could have been obtained from the Hellman-Feynman theorem, but no one was looking there.

My analytical derivatives treatment allowed me to develop a computational approach that treated π-electron molecules in the same way we treated non-conjugated molecules [14]. This advance did not sit well with Norman Allinger, the force-field guru of the organic chemistry community. Allinger developed programs that used the "bond order-bond length relationship", where the π electron bond or orders were calculated somewhat inconsistently in another program. When he learned about my approach, Allinger wrote me that he'd had a similar idea and had told Shnieor about it, who presumably told me, but that his student was drafted and went to Vietnam so nothing came of it. I responded politely that I never discussed quantum mechanics with Shnieor, and that I too had fought in a war.

I greatly enjoyed my time in Cambridge and developed significantly in terms of understanding my scientific strengths. I realized that one might have a lot of good ideas, but knowing which one works requires verification by computer simulations. I also began to understand that a good idea should be guided by a relevant question.

It is important to note that the CFF program that I brought with me from the Weizmann Institute has been adopted with some cleaning and modifications by Karplus' group under the name CHARMM, and has become a widely used program and the basis of most current molecular modeling programs (e.g., Amber, GROMACS and more).

In January 1972, we moved back to Israel, with a small amount of money that we'd saved from my small salary. That might have been for naught because on a layover in London we left the handbag with our money in a taxi. Fortunately, Tami spotted the taxi five minutes later and stopped it, the money still in the back seat.

VI

Back at The Weizmann Institute

1972–1973

I returned to the Weizmann Institute eager to start my career, getting immediately and atypically the rank of senior lecturer. I was not completely sure how my research should proceed; I considered several options, including some that could be viewed as extensions of my postdoc. My projects included an improvement of the CFF+π (QCFF/PI) and a new direction (QCFF/ALL) that was aimed at quantum mechanically treating all the valence electrons of the simulated active part. This would allow me to study actual bond breaking and would be implemented in a special hybrid orbital representation for future studies of enzymatic reactions. The QCFF/ALL treatment had the great advantage that the orbitals were pointing in the direction of the bonds (rather than in the x, y, z directions), but programing the relevant formulation was very challenging. I got significant help in cleaning the QCFF/PI and on developing the QCFF/ALL from my talented programmer — and in the years since, a close friend — Ruth Sharon.

I also started to develop a quantum mechanical treatment for simulating the electronic and vibrational states of molecular crystals in a model that I called "the rigid lattice approximation". I thought at the time that this could be very promising.

Sometime after my return I was happy to see that Mike Levitt joined the Lifson group again, after completing a groundbreaking PhD at the Medical Research Council that extended his protein modeling to many key systems, ranging from lysozyme to hemoglobin, showing the power of the energy refinement approach. I clearly looked forward to collaborating on studies of enzymes.

The Yom Kippur War: 1973

Toward the end of 1972, our 377th battalion was called up for a few weeks of reserve duty in the southern part of the Golan Heights. One day a very senior officer from the Northern Command came to the base, and one of our guys asked him what we should do if a Syrian division comes from the southeast. The response was very reassuring at the moment, but ominously misplaced: "Just send three of your tanks in that direction." His brash confidence was based on the IDF's experience from the Six-Day War, which unfortunately reflected a terrible misconception that would result in the devastation of the imminent Yom Kippur War.

I returned to the Weizmann Institute when my reserve service ended. I was immersed comfortably in my daily research until Yom Kippur Eve, October 5, 1973. Tami and I had gone to the synagogue and were surprised to see military Jeeps speeding far above the allowed limit, indicating that something serious was happening. The next day at 2 pm Israel was attacked by the Egyptian and Syrian armies and the Yom Kippur War erupted. I was not immediately called up because my unit didn't have my new address. When I finally got to the 377th base the battalion had already left, with the tanks moving on their chains on the asphalt roads instead of being moved by tank carriers. I didn't yet realize that the Syrian army, which completely outnumbered the regular Israeli army in the north, had already taken most of the Golan Heights and was somehow held back by the slowly arriving reserve forces. Mid-morning on October 8, I got to the road on the southern side of the Golan Heights, where I was stopped and not allowed to continue. From the radio in one of the parked jeeps I could hear the battle sounds of the Golani Brigade as it tried to recapture the crucial intelligence center on top of Mount Hermon, which had been taken by the Syrians on the first day. Unfortunately, the voices involved calls from the fighting units to the brigade command, announcing that most of the commanders had been killed or wounded. This battle would end badly. Later we were allowed to continue and I joined the 377th on the Golan Heights, where I was told that we were the only viable force on the southern part of the Golan front. This sounded unbelievable, and when a captured Syrian soldier told us they had taken the entire Golan, I thought he had lost his mind. On the morning of October 9, we started to move northeast; the battalion commander, Benzion Padan, told us that we were the only force in the region that separated the Syrians from the Jordan Valley and the rest of Israel. We moved forward as a part of a divisional attack aimed at pushing back the main Syrian army, whose command was located near the village of Hushniya. Although our super Sherman tanks were not exactly a match for the superior Syrian T55 and T62s, our 377th battalion moved to the center of the attack. At some stage, our command armored half-truck ran over an anti-tank landmine, one of the many spread on the ground overnight by the Syrians, and several of the command group, including Padan, were wounded. I jumped down and ran to the tank of a company commander, climbed up on its side, and told him to take command of the battalion. Unfortunately, I learned that my notification on the radio about our situation had sounded to some as though Padan had been killed, but after tending to his wounds, he returned on a jeep. Our strategy of head-on attack had not been so successful, and eventually the battle — by some accounts the most crucial in pushing back the main Syrian force — was decided by 15–19 tanks of our battalion that attacked the Syrians and led to their collapse and retreat.

Over the next few days, the Syrian lines along the post-1967 border were broken by the advancing Israeli army. However, on October 12 the overstretched Israeli column was subjected to a major threat when an Iraqi division supplemented by Jordanian forces moved from the southeast to cut it off. The general command was to retreat very fast back to the west. But, in part because our tanks did not have enough gas, Padan decided to stay and placed our tanks that did not have night vision on a back slope so they will see the Iraqi tanks on the horizon with the help of the moonlight. This led to an early morning ambush that centered around our outdated Sherman tanks, which nevertheless managed to destroy a significant part of the Iraqi forces. Although our commander got a special citation, the battalion's contribution remains largely missing from the history books. However, in the 1990s, a short time before his death, the division commander, Musa Peled, apologized for not emphasizing the role of the 377th battalion in the battle. And in 2018, when Nadav Padan, the son of our battalion commander Benzion Padan, became a major-general (*aluf*), the newspapers mentioned the past contributions of his father's unit.

After the ceasefire on October 25, the situation continued to be tense, with renewal of shelling and a war of attrition, until then-US Secretary of State Henry Kissinger brokered a permanent cessation of hostilities. At any rate, I was allowed to leave at some stage and I returned to the Weizmann Institute.

Protein Folding and an Early Enzyme Program

Returning from the war, I was in a bad mood and quite depressed. Nevertheless, I continued to work but decided to concentrate more on the biological track and stop the rigid lattice vibronic project. In particular, I focused on attempting to write a QCFF/ALL program for enzymatic reactions. In addition, I cleaned the program for modeling molecular crystals, together with Eduardo Huller and Ruth Sharon, adding the QCFF/PI quantum features to it, and also launched a project that studied crystalline excited state molecular pairs (excimers) [15] and the asymmetry in photochemical reactions of molecules in crystals [16]. One such study involved a collaboration with an experimental group that tried to determine the structure of excited crystals (in the excimer state) by X-ray diffraction. These projects advanced nicely.

One afternoon in early 1974, I sat in the computer room with Mike and we discussed how to address the protein folding problem. This challenge was considered by some to be the holy grail of molecular biology, since presumably if you could go from a sequence to a structure you would know how biological molecules work, as you will have the relevant structure. While I did not share this opinion, I realized that this was a problem of enormous "public" interest. One of us mentioned that if we had a ball-and-springs molecular model of a protein (with magnets serving as attraction centers), and if we could put it in a space without gravitation, we would be able to explore the problem.

Then one of us (I think it was Mike) suggested sending such a model in a spacecraft, where the gravitational force is almost zero. At that point I said that if we could replace the amino acid side chains by spheres, we would have a feasible way to study folding on a computer, since I knew that an all-atom model would not be able to tell us much with the computer power of that time. Mike was excited and immediately started writing the corresponding code. In short time, we got very encouraging initial results, where some minimization attempts led to folding. I argued, however, that our finding would not provide a major change in the understanding of protein functions. At that point, Mike said, "What do you mean, sharing half of the Nobel Prize will be great!" And this turned out eventually to be correct, although more due to the QM/MM work (see below) than because of the folding work.

Once we obtained several correct folding events of a small protein (BPTI) the excitement increased, since according to what has been called the "Leventhal Paradox" one could argue that it is very unlikely that a protein with so many degrees of freedoms would find a way to fold. However, averaging effectively over less relevant coordinates, we generated a model with significantly fewer degrees of freedom than a full atomistic model. With this reduced model we were able to show that the folding should not be formulated as a process that requires a complete search over all degrees of freedom and that folding can be simulated [17].

Interestingly, Shnieor said we shouldn't be so excited, since asking how a protein folds is like asking how a leaf falls off a tree. On the other hand, Mike felt that our approach would help in predicting protein structures. I was less optimistic, but felt that it would let us understand the mechanism of the folding process. Either way, since we had attacked what was seen as one of the most important problems in molecular biology, our work received significant publicity.

In March 1974 Tami gave birth to our second daughter, Yael, and this helped to improve my mood.

In late 1974 I decided to enhance my biology experience and to join Mike, who had returned to the Medical Research Council (MRC). My appointment as a European Molecular Biology (EMBO) Fellow was approved by MRC director Max Perutz.

VII

MRC, the Cathedral of Molecular Biology: 1974–1976

My wife, our now two daughters, Merav four and Yael one, and I arrived in Cambridge, England (Figure 14) in October 1974. We rented an apartment near the train station and I set to work.

The MRC was the Mecca of molecular biology, counting several Nobel Laureates on each floor. I moved to a relatively large room (the MRC generally had small office spaces) with Mike and my PhD student, Andrea Lappicirella. One of the neighboring offices hosted John Walker, who was working on ATPase, which led years later to his Nobel Prize.

Although my spirits were still low, I started to work intensively, taking up several directions. One was a continuation of the folding project, focusing on folding of helical proteins (where the helices were kept in the helical form during the folding process). In parallel, with Andrea I developed a QCFF/PI-type model for heme proteins, and for conjugated molecules with hetero atoms (atoms other than just carbon and hydrogen). This includes quantum treatment of the metal (e.g., iron) in the center of the heme group, considering explicitly the metal active d orbitals [18]. The model has been developed for treating hemoglobin (the protein that binds oxygen in the blood) and related systems. Eventually I used this model in exploring the proposed role of the metal spin states in the action of hemoglobin. In that case I was able to show that the change in the spin state is the result of the change in the heme structure due to the repulsion between the oxygen and the heme nitrogen atoms upon oxygen binding.

Figure 14.   1975, in Cambridge. Upper row from right to left: Tami, Arieh and Tova (Tami's sister). Lower row from right to left: Merav and Yael.

This finding contradicted the widely held assumption that the spin state changed due to the oxygen binding forcing the heme to be planar.

One of my main efforts was the attempt to model a reaction in a real enzyme (lysozyme). The first step was to try to represent the reaction of breaking a sugar glycosidic C–O bond quantum mechanically, and then to include the rest of the enzyme by a classical force field. To my surprise, the glycosidic bond "refused" to break, without investing more than 80 kcal/mol in the breaking process. This was strange since in the enzyme the bond-breaking energy is about 20 kcal/mol.

Taking this finding as a major guide to tackle the problems in my approach, I came to the realization that I had missed the most important factor in the effect of the enzyme (or solution) on the reacting bonds — the stabilization of the charges developed on the fragments of the broken bond.

In the gas phase, the bond is broken into two neutral radicals going uphill in energy by about 80 kcal/mol. However, in a polar environment the bond is broken into an ion pair that is stabilized by the solvent dipoles. Figuring out that this effect was not included in the QM+MM treatment was extremely instructive. In fact, this realization provided another demonstration of the power that can come from trying to model physical effects and failing to obtain the correct observed results.

Now, after discovering the main missing part in real QM/MM treatments, namely, the coupling to the environment, I had to figure out how to evaluate the electrostatic potential from the water and protein around the reacting bonds.

The first challenge was how to treat the effect of the potential from the charges of the environment around the quantum part. The main focus in the theoretical community has been on evaluating the interaction of the surrounding environment with the overall wave function of the quantum part (the solute). While this has led to interesting formal expressions, it was not highly useful. By going to the first step of the quantum derivation, I ultimately discovered that it is best to just introduce the electrostatic potential from the residual charges of the atoms in the environment in the solute Hamiltonian, by changing the effective ionization potential. This makes atoms that are subjected to a positive potential hold electrons much stronger than those that are subjected to a negative potential. This extremely simple treatment has been the basic of the QM/MM coupling (Figure 15). Interestingly, many years later I read a part of a lecture that argued that the QM/MM idea is trivial, and I eventually told the person who wrote the lecture notes that it is indeed trivial after you already know what the problem was and what treatment to use.

The next stage in the QM/MM treatment involved figuring out how to treat the protein/solvent environment (the generalized solvent) around the quantum region. After reading many books on classical electrostatics and talking to experts on electrostatic theories, I realized that the widely used continuum electrostatic description (where the environment is considered as a structureless medium) would not provide a satisfactory practical guide that could be used with confidence.

Apparently, even the basis for the famous factor of one half in the equation for the energy of an electric field (where the energy of a dipole in a field is only half of the product of the dipole and the field) has not

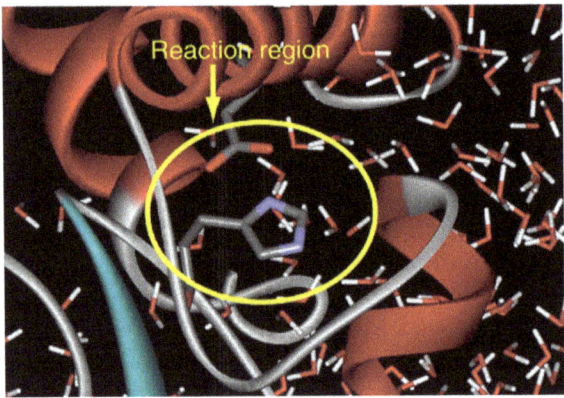

Figure 15.   A QM/MM model for the enzyme trypsin. The inner circle contains the part that is treated quantum mechanically, while the rest of the system is described by a classical force field. However, the electrostatic potential from the external part is included in the Hamiltonian of the quantum region. This multiscale approach allows one to study chemical processes in very large molecules.

been based on a molecular picture and was considered logically problematic in key textbooks (e.g., Bottcher) [19]. This was a major problem, not least in cases where trusting the results was paramount. Thus, I concluded that I must move to an explicit (yet simplified) representation of the surrounding environment.

This began with moving to a polarizable force field for the protein atoms, where each dipole, assigned to a protein atom, responds to the field from the protein charges (times the atomic polarizability) and to the self-consistent field from other polarizable dipoles. In this way, we didn't need to use any phenomenological dielectric constant for the protein interior (eventually we also had to include the microscopic protein relaxation). As a side point, representing the atomic electrons as polarizable dipoles resolved the issue of the factor of one half in the field energy [20].

The second focus was the crucial treatment of the water molecules in and around the protein. This was done by representing the water with a grid of Langevin-type dipoles (which are polarized according to the average thermal distribution). My point was that if we somehow knew the exact solvent polarization around charges in water, we would have known the correct solvation energy of the charges (the interaction between the charge and the dipoles, plus the energy of orienting the dipoles). The idea was to choose empirically a charge-dipole dielectric function that would reproduce the observed solvation and thus generate the correct polarization. Incidentally, I was asked by David Blow, who was the first to solve the X-ray structure of chymotrypsin, to give a short lecture on this water model in a Faraday discussion meeting, but unfortunately no one appeared to follow my inverted logic of using the observed solvation to calibrate the polarization. The fact that the water system was described by a simple dipole-based model led subsequently to ceaseless criticism. For example, it was argued that it is incorrect to use dipoles when everyone knows that a water molecule is not a dipole. This argument, of course, missed the fact that our dipoles were "effective" and were calibrated based on solvation free energies. Actually, for about ten years, we were arguably the only group with a clear physical understanding of solvation and electrostatics in proteins. This was the case since we insisted on using a simple but complete model for studying the solvation of charges in proteins, while the rest of the community was satisfied with either continuum models that used an undefined dielectric for the protein, or with calculations that considered a single ion or a single HCl molecule as a guide for studying solvation effects.

I also made another major advance by modeling the initial photochemical reaction in the visual process. This process starts when light is absorbed by the protein rhodopsin in the eye and excites a retinal molecule that is bound to the protein as a Schiff base. Upon excitation the retinal rotates around its 11-12 bond and is transformed from a *cis* to *trans* configuration and then returns to the ground state in a *trans* configuration. The isomerized retinal pushes the protein to a new configuration and activates the G-protein transducin on the other side of the membrane, leading eventually to a transfer of information to the brain. Knowing about the importance of the visual process I became very interested in its primary event and the initial photoisomerization process.

In 1971, after realizing that I could model photochemistry by a semiclassical surface-hopping approach (see chapter V), I was extremely eager to advance this idea. However, Karplus, who did appreciate the semiclassical direction, was still hesitant about moving forward. In 1972, I wrote a paper that described the method and explored the photoisomerization of butene-2, but the paper was published with Karplus only in 1975 [21] (and again in 2014 after the Nobel Prize). Nonetheless, Karplus probably talked about this idea with other scientists as I was told in 1974, and in 1978 I even saw that a version had made the cover of *Modern Molecular Photochemistry* by Nicholas Turro. The engraved image was incorrect, however, since the trajectories at the crossing point split into two directions instead of retaining momentum and moving in one direction. Still, this all indicated to me that I might lose primacy on my important idea. So in late 1974, I decided I shouldn't wait anymore and started working on simulating the primary event in vision. It was clear to me that elucidating what actually happened after absorption of light by the eye would be an extremely important advance in exploiting the potential of computational chemistry in life science, and I was eager to be the first to perform such a study. I used the QCFF/PI potential surface and simulated the surface-crossing dynamics of a protonated Schiff base of retinal subjected to a restraint that modeled the constraint of the rhodopsin active site. This simulation intended to reveal what actually happened in the rhodopsin protein in the eye at the first picosecond of the absorption of light.

I remember discussing the validity of my surface-hopping approach, which ignored interference between different trajectories, with Bill Miller from the University of California at Berkeley, who was on sabbatical in Cambridge. Our meeting was in the cafeteria at the old Cavendish Lab building, and although Bill was the world expert in this field, he didn't have a clear opinion on my intuitively reasonable approach. I went forward anyhow and completed what was the first molecular dynamics (MD) simulation of a biological process [22]. I obtained a very fast photoisomerization time of about 100 fs (Figure 16), which was much quicker than the lower limit of 6 ps observed at that time. Miraculously, my simulation time predicted the correct experimental value that was measured several years later.

This success was possible because the focus of the simulation was on an extremely fast process that could easily be simulated using the computers of the mid-70s. I also evaluated the surface-crossing probability and reproduced the exceptionally large observed quantum yield. This reflected the fact that my surface-crossing treatment found by itself the so-called

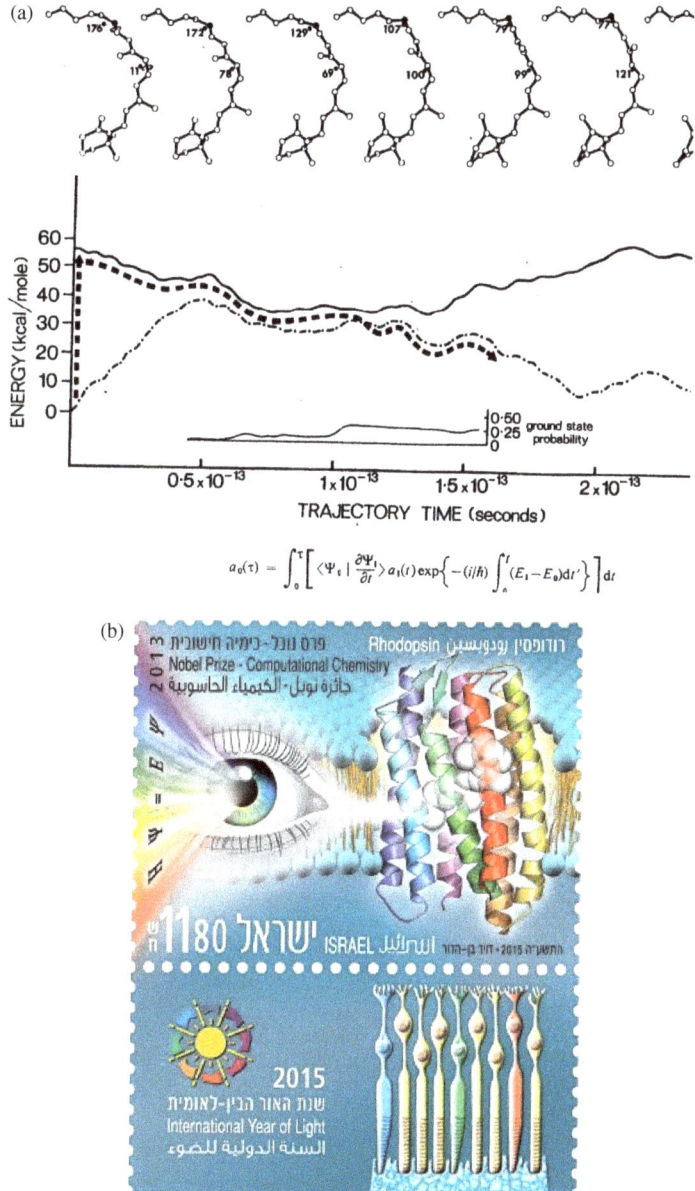

$$a_0(\tau) = \int_0^\tau \left[\langle \Psi_1 | \frac{\partial \Psi_1}{\partial t} \rangle a_1(t) \exp\left\{ -(i/\hbar) \int_0^t (E_1 - E_0) dt' \right\} \right] dt$$

Figure 16. (a) The results of the 1975 MD simulation of the primary event of the visual process (taken from [22]). The figure describes an excited-state trajectory in a model of a protonated Schiff base with an artificial constraint that represents the restriction imposed by the protein. This simulation predicted correctly the photoisomerization time and the observed quantum yield. (b) An Israeli stamp issued for the Year of the Light and our Nobel award, featuring the simulation of the primary event in the vision process.

"conical intersection" of the system where the transition between electronic surfaces is very effective. This occurred long before I knew what a conical intersection was, and long before the subsequent discoveries that conical intersections play a major role in most photochemical processes (e.g., [23]). This demonstrated the power of computer simulations in discovering new phenomena, without a complete realization of the underlining theoretical background.

Since the structure of rhodopsin was not known at that time (and not for more than another 25 years), I thought it would be a good idea to bind retinal as a Schiff base to chymotrypsin and to examine the photochemistry of this system, and asked Richard Henderson to perform the experiments. Richard was involved at the time with electron microscopy of bacteriorhodopsin, and many years later received the Nobel Prize, but he replied that during his PhD, under David Blow at the MRC, when he was crystalizing chymotrypsin, he acquired an allergy to the protein.

Cambridge was a mix of work and family. I would take Merav to school, walking with her over the bridge near our house, and then head to work. I'd come back for dinner, play with the girls and tell them stories when I put them to bed, including the beloved Barmeli tale that my father used to read to me. Then around 9 pm I went back to the center, trying to submit one more set of jobs to the computer. I stayed home once a week so Tami could go to her different activities, like wheat weaving and copper enameling. On weekends we traveled around England, visiting London and sites around Cambridge. Yael would be carsick on every ride, losing her food after about three miles but happily recovering after that. Despite that one slight hitch, I greatly enjoyed these trips.

Back at the MRC, I had a productive interaction with Max Perutz, the father of structural molecular biology. On the one hand, I was slightly intimidated by Max, who had little patience for discussions that did not interest him. But I was trying to understand what the main issues were in describing the allosteric action of hemoglobin. Our initial discussions were short but helpful and when I left the center, Max sent my papers to the *Proceedings of the National Academy of Science (PNAS)* [18] and wrote a major review in *Science* on electrostatic effects in biology that kindly featured my work. In 1984, he asked me to write a *PNAS* paper that would explain the problems with the Tanford-Kirkwood model and other continuum electrostatic models that considered the protein as a low-dielectric oil drop.

I also had some interesting dealings at MRC with Aaron Klug, who was later awarded the 1982 Nobel Prize for his structural studies of nucleic acid-protein complexes. Before he arrived at the center, Aaron had worked in J.D. Bernal's department at Birkbeck College and was somewhat obsessed with great symmetries, like those found in the coat proteins of viruses. He grew up on the idea that water solution has clusters of water molecules with the structure of clathrates, and so he had problems with my attempts to model solvation by a grid of dipoles. I tried to make my point by showing him that if we take one large coin and attempt to surround it by smaller coins, we will never get a perfect packing, hence the water around ions of different size cannot have a perfect structure, but I don't think that this argument convinced him.

Near the end of my time at the MRC, we had a small meeting in the Netherlands, which was largely inspired by my and Levitt's works. The organizers were interested in modeling the action and dynamics of proteins and suggested having a follow-up workshop. Unfortunately, in 1976, when the workshop was ultimately held, I had already become a professor at the University of Southern California (USC) and felt I would be too busy and did not attend. I came to regret not participating, since this workshop has been frequently featured as the origin of simulations of biological dynamics despite my vision work.

At that time, I knew that modeling the dynamics of proteins with the available computational resources was completely useless, since no convergence could be obtained. But I overlooked the benefit of being able to claim I had done something first, even if the study was not really relevant. In this respect, if I had attended the workshop I would have most probably repeated my MD simulations of the first step in vision in an artificial site of some protein. This would allow me to claim that not only had I done the first MD simulation in a biological system (the vision work), but I also conducted the first relevant MD simulation of a process in a protein. This was particularly important after the editors of *Nature* changed the proofs of my vison paper, modifying the initial title from "Molecular dynamics simulation of the vision process — bicycle pedal model of the first step of the vision process" to "Bicycle pedal model of the first step of the vision process". The editor claimed that the word "dynamics" did not exist and that the title was too long. In fact, for a long time I thought these two episodes significantly reduced my chances of receiving a Nobel Prize.

A noteworthy process that occurred near the end of my time at the MRC was a major change in my perspective: I almost suddenly started to see biological problems as clear problems in physics, rather than fuzzy phenomenological problems. I stopped being focused on reproducing experimental results not directly related to the given function, and turned my attention to reproducing the overall process, like rate constants. I also stopped accepting the experimentalists' explanation of the action of a biological system as a clear guide. I am not sure when and how this mental switch began, but looking back it was a crucial transformation.

Along these lines, I should mention that back during my PhD and sometime after it, I assumed that focusing on vibrational normal modes could provide considerable information about molecular properties. This was the implicit assumption of many chemical physicists, who thought that the harmonic force field should give a powerful description of molecular behavior. Thus, I put substantial effort into developing methods for evaluating infra-red and resonance Raman spectra (e.g., [24]). However, eventually I realized that while this information is important, evaluating the whole energy landscape of the given process is what is needed from computer simulations of biological systems. In other words I realized that in order to understand biological systems I must be able to simulate the actual relevant processes rather than individual properties (such as structural features). This realization became my guide in my future advances.

VIII

USC: 1976–1981

At the end of 1975 I decide to consider staying in an American university at a tenure track, while seeing the progress of my tenure process at the Weizmann Institute. I traveled to the US for several interviews and received a positive response from the University of Southern California (USC), where Otto Schnepp, with whom I had worked at Weizmann while he was on sabbatical, was a faculty member, and where Gerald Segal, whom I met at the Weizmann in 1972, was very interested in my work on photochemistry. I accepted the USC offer and eventually got a Green Card in the US embassy in London, with the help of an ex-USC faculty who served as a cultural attaché to England.

Subsequently we arrived in Los Angeles in February 1976 after a long flight from London, on which my younger daughter, Yael, one month shy of being two, did not stop crying and the passenger behind us did not stop trying to stop the crying by kicking our chairs. We were welcomed in LA by two extremely rainy weeks. Fortunately, Otto Schnepp allowed us to stay in his condo, La Ronda. I joined the USC chemistry faculty and was assigned to teach Group Theory, which was not exactly my strongest point. I read several textbooks and usually was able to teach the material that I had learned the previous night.

In the summer of 1976, I attended my first Gordon Research Conference, where I tried to describe the QM/MM lysozyme work. That was also the year I regrettably declined the invitation to that key workshop in France, first planned in the Netherlands in 1975, about biomolecular simulations. Another fateful event occurred that summer, when I received a letter from Shneior telling me that he did not have good news for me — my tenure at the Weizmann Institute was not approved, although I was promoted to associate professor. He pointed out that the reason was a recommendation letter stating that while I am very bright, it is impossible to know whether to believe my findings. I was told off the record that this was written by someone whom I worked with and knew very well that I never made up my results. The only poetic justice in this story was that the tenure committee had examined the manuscript of my 1976 QM/MM paper [20], which was at the heart of my 2013 Nobel Prize. In fact, I was told by my friend Atilla Szabo of the National Institute of Health that I probably am the only case of a Laureate awarded a Nobel based on a paper involved in his tenure denial.

So I remained at USC — not a bad thing, it turned out. At the end of spring semester in 1977, I accepted an invitation to help Kent Wilson from the University of California, San Diego (UCSD) during the summer in his attempt to develop a machine language processor that would perform force field calculations of proteins. Wilson's progress with his processors eventually become a reality and almost led to commercial machines in the early 80s, but this came at the same time as regular computers became faster and cheaper. I enjoyed the summer and while I was there, started to work on free-energy perturbation study of ionic solvation, using a soft sphere dipole as a water model. These simulations converged very slowly, and regretfully I never published the results. Thus, I missed bragging rights, this time for being the first to perform solvation free-energy calculations. I still got this notice in a 1982 paper, but as a side point in simulations of electron transfer in solution.

As to the solvation calculations, I mainly focused on an energy minimization study of the spherical dipole model, which captured the solvation physics very effectively and converged in a reasonable way. My attempts to publish this model, starting at the end of 1977, led to enormous resistance and was rejected by several journals. The referees' criticisms were associated with many years of experimental focus on second-order factors in ionic solvation, which actually involves enormous energy much larger than the small variation due, for example, to changing temperature. Most of the attention at that time was dedicated to the special effect of the structure of water molecules. However, I concluded that for developing a way to evaluate solvation-free energies, the relative difference between solvation in methanol and in water is trivial, and I could take a dipolar model and parametrize it to have some feature of hydrogen bonding and reproduce solvation energies in water. I had to face the "well-known" counter-argument that you cannot model water by dipoles, but this claim, as noted above, overlooked the fact that it is related to modeling a water molecule by a dipole with the actual water dipole, and not to the trivial ability to have a dipolar model calibrated to reproduce the solvation properties of water solution.

Illustrating this point, a leading chemical physicist wrote in his review of my paper: "How can one model water with dipoles, neglecting quadrupole, when everyone knows that the solvation of argon is so much different from the solvation of the isoelectronic Cl^- ion." This problematic argument underscores how some in the community did not realize that the solvation of ions is very different to that of non-polar solutes. This was a clear example of focusing on exotic but unimportant features. One of the most telling rejections included the statement that my model does not provide a "complete Hamiltonian". I never fully understood this misguided assertion, since my model included the whole system up to infinity, where the soft sphere dipoles were subjected to surface constraints and surrounded by bulk. However, the fact that the system was built in a coarse-grained way was considered unacceptable by researchers trained in rigorous descriptions, and who gladly accepted quantum mechanical calculations with one HCl molecule and a Cl^- ion as a major advance in understanding solvation effects.

Problems with the publication of the dipolar water model continued until the end of 1978, when I was invited for the first time to be a plenary speaker at a major conference, the International Biophysical Conference in Kyoto, to speak on the origin of enzyme catalysis. Professor William Lipscomb, who attended the lecture, subsequently invited me, along with other much more distinguished people, including Aaron Klug from the MRC and Tom Steitz from Yale University, to participate in a major event in Minnesota, which was convened to underscore that Lipscomb's Nobel Prize work was done there.

The story now goes back to my solvation paper that was at the rejection stage from the third journal, *The Journal of Physical Chemistry*, with the unhelpful comment that I had not calculated entropic effects. I responded that the entropic contributions to ion solvation are extremely small relative to the total solvation free energy, and that I can even estimate such effects by free energy perturbation and consider the minimized calibrated energy as an approximation for the solvation free energy. And I wrote to the editor, who was from the University of Minnesota, that unless he accepts the paper, I will make my entire enzyme lecture at the upcoming conference about these referee reports and how their focus on trivia will prevent a real understanding of enzyme catalysis. This led to the article's speedy publication [25].

The saga of my soft sphere dipole paper marked my graduation into the world of negative and sometimes even hostile reviewers, which I had not experienced until that point and never left afterward. One explanation was provided by the associate editor of *The Journal of the American Chemical Society (JACS)* in the first round of the paper. This very distinguished physical organic chemist wrote that I should talk with the readers and not at them. The problem with this proposal has been that people who have focused on specific types of effects, e.g., measuring small effects of various perturbations, were not inclined to listen to those who concentrate on different effects, such as much larger absolute contributions, like solvation effects, which are not determined by direct measurements but are the main factors in the overall energy.

A single encouraging example of the problems in grasping my direction occurred in 1977 when I gave a seminar on enzymes at UCSD. Generally, I got little response and a few relevant questions. However, at some point an elderly man raised his hand and said, "Yes, but we did not have computers," implying that if he had a computer he would have done the same. I found out later that this was J. E. Mayer, the great statistical mechanician credited with the first attempt to develop a force field. Hopefully he realized that I had made a breakthrough by using computers.

In 1978 Max Perutz published what became a very influential article in *Science*, "Electrostatic Energies in Proteins" [26], which heavily featured my work, emphasizing my idea that enzyme catalysis is due to electrostatic effects. It also implied that the catalysis is attributable to having charges in a low dielectric. In this case he overlooked my non-trivial point which was that in proteins charges are stabilized in very polar but preorganized environments. But overall it was very supportive.

In the summers of 1979 and 1980, I spent time in Del Mar, San Diego, where the main purpose, aside from vacation with my family, was to collaborate with Mike, who was then at the Salk Institute. We were trying to write a general book on the fundamentals of biomolecular modeling. We started with outlines and then moved on slowly with different chapters.

These were very nice summer vacations. We stayed at a beach motel and met every afternoon after my workday on the San Diego beach, ate sandwiches and tried to swim in the very cold water of the Pacific Ocean. Some mornings we also tried venturing into the ocean, and in the evenings we sometimes lit a campfire on the sand and sat around it. We were joined occasionally by the family of one of the premier jockeys at that time, who were staying in the same motel during his races at the nearby Del Mar racetrack.

At this point, I started to attend major conferences, trying to convince people of my ideas, in the hope that direct presentations would be more effective than merely publishing papers. Another useful experience involved explaining my study of the first step in the visual process. In 1980 my postdoc Bob Weiss [27], animation guru Mallory Pearce and I created a movie in a Hollywood animation studio, based on my 1975 simulations. One of the first films in the field, it depicted the process of absorption of light and the subsequent MD trajectory (see Figure 16a) of the chromophore in rhodopsin. At a Gordon conference on vision that same year, a friend invited me to show the movie at a nearby conference on protein crystallography, and I was eager to oblige. When the movie ended, however, a crystallographer raised his hand and asked, "What evidence do you have for this?" I was completely shocked, since the results reflected about five years of research, and was to eventually exactly predict the experimental facts. For him, it appeared, this was all just a fictitious cartoon and he was reflecting a complete disbelief in computer simulations. I could have replied that I had far more evidence than he ever obtained for any of his mechanistic proposals, but held my tongue, knowing this wouldn't help advance the importance of computer simulations.

I had a similar experience at another conference, where I tried to explain to Dan Koshland, a guru in the field of enzyme catalysis, my 1976 QM/MM papers and other ideas I had been working on. He seemed interested, but Shneior later told me that when he asked Koshland what he thought of my work, he responded, "I have no clue what he's talking about." Interestingly, Koshland proposed that enzyme catalysis is due to "orbital steering" [28], where the presumed need for exact orbital overlap leads to an enormous entropic contribution. This popular proposal, which was initially supported by very crude quantum mechanical calculations, has been shown to be completely unrealistic.

A related event occurred in 1979, at a conference in honor of Shneior's 65th birthday held at the Weizmann Institute. I gave a lecture on enzyme catalysis, and when I spoke about lysozyme I was attacked by the great crystallographer Jack Dunitz, who stopped me to deride my work, declaring that "there are countless works on sugar hydrolysis," which presumably I am not aware of. I responded in defense, "Yes, but none of them estimated the activation barrier, they only provided the rate constants." Later I heard from Ruth Sharon who had sat near Ora Kedem, a leading figure at Weizmann, that Kedem had told her, "He's bright but now I understand why he didn't get tenure."

In 1979 I got a position of Associate Professor with tenure at USC and had less pressure about my future. Overall I continue to progress in quantifying reactions in enzymes and solutions, developing with Wiess the Empirical Valence Bond method [47] that will be discussed later. This method provided a powerful way for quantifying my concepts.

IX

Sabbatical at the Weizmann Institute: 1981–1982

In 1981 I took a sabbatical at the Weizmann Institute as a senior EMBO fellow. My hope was to continue writing the book with Mike and perhaps attempt to return to Weizmann. We rented a nice apartment and both my daughters went to a regular Israeli school, not a simple feat. When Merav got a bad grade in Bible studies, for instance, she blamed it on Tami, who helped her with homework, saying that it was she who had done poorly. Otherwise, our parents were very happy to see their grand-daughters and we enjoyed traveling on the weekends.

I began polishing my article on the dynamics of reactions in water [29], which I delivered in the summer of 1981 at a theoretical chemistry meeting in Boulder. This paper provided the first microscopic treatment of the dynamics of electron and proton transfer in condensed phases. Most significantly, it presented the first microscopic evaluation of the Marcus parabolas for electron transfer reactions, which Rudolph Marcus had proposed based on continuum considerations that led to his Nobel Prize in 1992. It was the first free energy perturbation calculation of a charging process and the first surface hopping approach for electron and proton transfer reactions. These were major advances, in particular the work on electron transfer. However, it took almost a decade and several additional papers until these works began to be adopted.

It is interesting to note that despite the enormous contribution of Marcus' electron transfer theory, most scientists at the time did not really understand the physics of this process, which had been presented by a complex macroscopic derivation without a clear molecular description (the theory's acceptance was based on its prediction, which led to Marcus' well-deserved 1992 Nobel Prize). On the other hand, my microscopic model was based on a clearly defined model of fluctuating solvent dipoles and on a reaction coordinate that involved the energy gap between the potential energies of the reactant and product states. This model allowed me, and others, to clearly comprehend the meaning of free energy parabolas and how they are related to the possibility of reaching the point where the electron transfer occurs. However, even my microscopic model with fluctuating dipoles that change the relative energies of the reactant and product states was not understood for a long time by a good part of the community. The best example of this was an electron transfer meeting in Philadelphia where I presented a film illustrating the effect of fluctuating dipoles on the electron transfer process. Barry Honig later told me that in the restroom after the movie he overheard two distinguished scientists saying, "What are those stupid dipoles Warshel is showing?"

In 1981 I summarized my conclusions, in *Accounts of Chemical Research*, that electrostatic effects are the key to catalysis and other biological actions, which was picked up by the journal *The New Scientist*, and sent to me by the academic secretary of the Weizmann Institute. I thought that it would make a real impression, but the immediate impact was not so great. Mike and I also met several times a week to push ahead on our book. We made some progress on various chapters, but overall didn't move substantially forward.

Near the very end of my sabbatical, on June 5, 1982, Beni, my youngest brother, got married, but was called up the next day for reserve duty for Operation Peace for Galilee, in what became the First Lebanon War. I didn't get a call-up order, as I was now based in LA. And soon after we headed back to the US, spending about two weeks traveling through Europe. It was a very enjoyable trip, even if Merav and Yael, now 12 and 8 respectively, sometimes fought in the car and didn't seem to appreciate the landscape and historical monuments. Incidentally, we had to buy a new return ticket to LA because Freddie Laker and his Laker Airlines, which we were set to fly, went bankrupt just at that time.

X

At USC: 1982–1990

Establishing the Validity of my Electron Transfer Simulations

On my return to campus, I planned to resume my work on biological problems, but I somehow again became involved in a major controversy. It began when I asked USC Chemical Physics Professor Howard Taylor about my idea of adding the radiation field of the absorbed light to my surface hopping model. He became extremely upset, screaming that my treatment of electron transfer in solution was all wrong, since I ran trajectories on the ground state, while Rick Heller's wave pocket treatment [30] involves running trajectories on the excited state. It didn't help when I explained that when I deal with excited states I run trajectories on the excited state and that both treatments are approximations that work exactly only in the harmonic case.

This was a typical case of believing fanatically in something that started with a rigorous derivation and was then followed by major approximations, rather than an intuition-based solid approach that starts with a reasonable approximated model. Taylor launched a major campaign against my approach, maintaining that I was completely wrong. He focused on one of my most important ideas, electron transfer as a surface hopping process, which later became widely used. I tried to explain my considerations to other faculty members, keeping in mind upcoming promotion to a full professor, but they basically had no clue about surface hopping and related approximations. Thus, as I frequently found to be necessary, I had to try and convince the chemical physics community that my approximations were as reasonable as the ones they believed in, and that none of these approximations were perfect. This started with proving that Heller's seemingly exact wave pocket treatment works exactly only in the harmonic case, and the same is true for my approach. That is, according to the Feynman path integral formulation, the rigorous treatment requires running trajectories at all possible energies, which is impossible in terms of computational resources. However, in the harmonic case, running at one energy has the same result as running at all energies (which means that we obtain the correct result automatically). I then developed, with the help of my terrific student Jenn Kang Hwang, a Green function trajectory approach that gave the correct anharmonic quantization [31, 32], which no one in the field had yet done. Subsequently, we were able to go from a rigorous Green function quantization to my semiclassical electron transfer approach, and show that the neglect of interference between the path integral trajectories [33] leads to the Marcus expression. Significantly, such a fundamental analysis had not been provided before.

Unfortunately, very few people followed or understood our Green function derivation [32]. Moreover, these studies "cost" me at least one year of serious theoretical work (we got only two reprint requests). And of course Taylor never changed his mind, which was problematic for him after my electron transfer model became so widely adopted. Even though the formal work slowed me down in my biological projects, it positioned me in the frontiers of theoretical chemical physics and helped me greatly in establishing my future formulations and defending my various approximations.

This whole journey reminds me of the story about Erwin Schrödinger, who was told by distinguished physicists, including Wolfgang Pauli, that his wave mechanics theory was problematic, while Paul Dirac's was the correct one. Although Schrödinger's approach worked in many more cases, he spent great effort showing that both were correct. When he ultimately proved successful, Pauli simply told him that he had always known it.

In 1982–1984, working on formal derivations, I did not sense that much was happening on the biological front. My colleagues in the field had not yet grasped the nature of electrostatic energies in proteins, focusing instead on a problematic continuum model with a non-polar protein interior, and they had still not tried to model enzymatic actions.

I even spent some of the money from my NIH vision grant on an experimental project trying to design an artificial light-induced energy storage molecule that exploits photoisomerization. But in late 1983 I started to feel that I had spent too much time justifying my surface crossing approaches. I decided to turn back to the issue of electrostatics in proteins, and started to write, together with my first PhD student Steve Russell, a decisive paper on electrostatics in biological systems. It clarified key relationships between microscopic and macroscopic dielectrics and covered all aspects of protein electrostatics [34]. It was in some respects far ahead of its time, since although more and more scientists accepted the continuum formulation based on the formally correct Maxwell equations, they unfortunately relied on a completely undefined dielectric constant. In fact, it took about 20 years until my views on electrostatics in proteins began to be widely accepted and supported by clear experiments. At that late stage, many researchers moved to fully microscopic simulations and did not appreciate the power of understanding and using a proper dielectric constant. In retrospect, this review should have been written as a book and I regret not doing so.

By this point, I had fully consolidated my view on the future direction of my research and decided that from now on I would focus on important biological problems, rather than on those that everyone else was working on. When I told this to a friend, he said that no one got ahead by solving problems. This had some truth to it, since solving a problem leads, in the early stages, to major objections from those with different ideas. Developing methods might have a larger impact if the method is widely used. However, in my view, there is more satisfaction in solving problems.

An interesting invitation came in 1983 from the Pontifical Academy of Science in the Vatican (founded by Galileo) to attend a high-profile meeting on the subject of biological specificity, which included top-tier scientists. I used this opportunity to submit a paper to the academy publication [35], and provided the first MD simulation and free energy perturbation calculation of an enzymatic reaction. The highlight of the

Figure 17.   Pope John Paul II and Tami in the Vatican during the pontifical academy meeting in 1983.

meeting was a lecture by Pope John Paul II (Figure 17), who discussed the relationship between science and religion, in which he said that the Church has no problem with science, but some problems like consciousness would never be elucidated by science.

After the lecture, a journalist asked me my thoughts, and I responded that some Jewish scientists did not see any contradiction between the Bible and scientific findings. Their answer to the question of reconciling the biblical contention that the world was created only 5,000 years ago is that no one knows what God considers a year. That got quoted in some journal and the story could have ended there, but about half a year later I was asked to give a seminar in a college in the Midwest. When I asked what the subject should be, I was told "the relationship between the Catholic Church and science". I had to disappoint them, and replied that I am actually a non-observant Jew and I wouldn't be a good speaker on that topic.

From 1984 to 1986 I also advanced an approach I called the "dispersed polaron" model [33]. We used Kubo's harmonic quantum mechanical formulation [36] for transition between electronic surfaces to obtain the quantum mechanical electron transfer rate constant at any temperature. The idea was to use the Kubo formula (which is applicable to harmonic cases) and obtain the needed Franck-Condon vibrational overlap terms from high temperature trajectories (where the classical and quantum results are similar). Then the quantum mechanical vibrational overlap could be used at any temperature in the Kubo quantum mechanical formulation. The basic idea for this treatment emerged from a related treatment of vibrational line shape that Shaul Mukamel developed for our work [37]. At a later stage I realized this idea's great potential for studying the temperature dependance of electron transfer reactions.

Elucidating the Nature of the First Step in Photosynthesis

One of my most exciting endeavors involved the race for understanding the details of the primary event in photosynthesis, among the most important processes in nature. It involves conversion of light energy to chemical energy, which is used to fuel the action of plants and certain other organisms. The light is absorbed by a molecular complex that includes a membrane protein with special light-absorbing chromophores. The excitation of the chromophores is followed by electron transfer and charge separation that initially stores the light energy as electrostatic energy [38]. The primary charge separation process involves a very fast unidirectional electron transfer, whose understanding is of great fundamental and practical importance.

I started to attack the challenge from several theoretical directions. Starting in 1979, I modeled the primary electron donor (the chlorophyll dimer), first reproducing its large absorption red shift, showing that it is due to excimer formation. I then explored the role of the entire chlorophyll dimer in the electron transfer process by evaluating all the vibrational levels and vibrational transitions between the ground and excited state of the dimer. This study showed that, in contrast to the common assumption at that time, it is impossible to block the back reaction by the effect of the so-called Marcus inverted region. That is, I found that the coupling between the ground and excited intramolecular vibrations allows the system to cross the corresponding Marcus classical barrier. Thus, the only way to have high efficiency is to have a very fast forward electron transfer. My paper [39] with these findings was communicated to *PNAS* by Martin Kamen, one of my heroes. Kamen together with Sam Ruben discovered the carbon isotope C14 and missed the Nobel Prize in part because he was falsely accused of giving secrets to Russian agents and was fired in 1944 from Berkeley. It took him a long time to obtain a job in his field and regain his good name. He was an extremely influential scientist and a very wise person.

Next, I was ready to explore the actual electron transfer in the protein. In 1984 I heard about the dramatic advances in getting the crystal structure of the bacterial reaction center at the Max Planck Institute in Munich by Hartmut Michel, Hans Deisenhofer and Robert Huber. I decided to make a pilgrimage to the institute and met with Hartmut and Hans, who showed me the initial model, which described the structure of the chromophores and the protein, but still did not include the assignment of the protein sequence. I asked Huber, who was the institute director, if I could give a seminar on electron transfer in proteins and photosynthesis. He agreed, but said dismissively, "Of course, Warshel can give a seminar on anything," not believing I was actually a leading expert on modeling electron transfer in proteins (being basically the only one who conducted such studies in this time).

In the lecture, I noted that until the discovery of the protein structure with the actual sequence, and the corresponding electrostatic potential from the protein on the chromophores, we would not be able to understand photosynthesis. This sounded strange to those attending, as they assumed, like most people, that just knowing the chromophores' positions should provide understanding of the primary event. Perhaps they also thought this issue could be completely resolved experimentally. Following the seminar, I proposed we collaborate and asked permission to get the coordinates, but they declined politely and said that the structure-function elucidation was a German project that they would keep to themselves.

I decided to continue with the preparation work, using only the positions of the chromophores. I formed a collaboration with my good friend Bill Parson of the University of Washington, a leading experimentalist in the spectroscopy and ultrafast kinetics of reaction centers, who had started to be very active and insightful with simulation projects. We started by evaluating the complex absorption spectra of the reaction center [40]. We also calculated the approximate electronic couplings between the chromophores [41], knowing that these couplings are needed for evaluating the rates of the electron transfer reactions. Of course, the big challenge was the analysis of the steps in the primary charge separation in photosynthesis.

As background for describing our analysis, note that the bacterial reaction center includes a chlorophyll dimer (P), which is the primary electron donor (see Figure 18). This pair is adjacent to two branches, each of which includes two chlorophyll-like molecules (the first is a

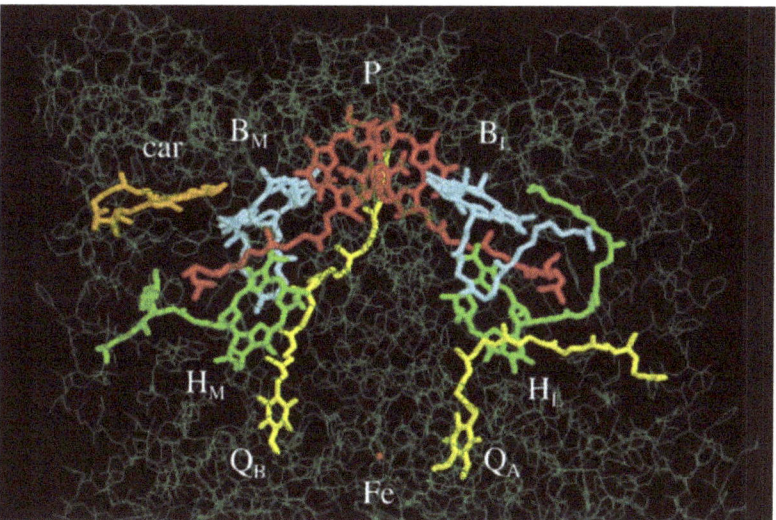

Figure 18.   The bacterial reaction center with the key chromophores: the special pair (P), the accessory BChls (B_L and B_M), the BPhs (H_L and H_M) and the quinones (Q_A and Q_B). Absorption of light in P initiates an ultrafast electron transfer that leads in about 3 ps to P^+H^-. The puzzle was whether or not we have a P^+B^- intermediate. The figure is taken from [38].

bacteriochlorophyll, B, and the second is a bacteriopheophytin, H). Each of these branches leads to a quinone electron acceptor. By 1986 it was well known that the P^+H^- charge separation occurs after absorption of light and takes about 3 picoseconds. However, it was entirely unclear if this process involves a two-step mechanism with a direct hopping (first P to B and then B to H), or what is called "superexchange", where B is at higher energy than P (in this case B is simply used to increase the coupling between P and H). Most researchers tended to believe in the superexchange mechanism, since the time-resolved studies did not show any intermediate before the formation of P^+H^-. However, we felt that it was essential to be able to evaluate the energetics of the different charge transfer states before making any meaningful analysis. This, of course, required the availability of the protein structure, which at that time, in early 1987, we still did not have. So it was exasperating to hear that the German group was collaborating with Klaus Schulten and Martin Karplus on calculations of the function of the system. I frankly was not worried about this collaboration, since neither Martin nor Klaus had any experience in consistent calculations of redox potentials or electron transfer, but at some point I heard that Klaus was trying to learn from Parson how to calibrate the redox-related parameters. Now in a panic about others getting the coordinates, we contacted George Feher at UCSD, who had a structure of a reaction center from a related bacterial species (*Rhodobacter sphaeroides*), but he declined our request. Fortunately, we found out that Marianne Schiffer and Jim Norris from Argonne National Lab were also progressing with a structure of the *sphaeroides* reaction center and convinced Jim to collaborate. Now we were cooking. We were the only group in the world with all the tools ready and tested. This included free energy perturbation calculations of redox potential, surface hopping calculations of electron transfer rates, calculations of electron transfer coupling and more. With great help from a very talented PhD student, Steven Creighton, we finished all the calculations in two weeks and submitted a paper to *Biochemistry*, where Parson was an associate editor. Parson chose not to apply his clout, and allowed a referee, who asked questions about the X-ray-observed water molecule, to delay the paper until January 1988 [42]. And still, this landmark article was the first to correctly determine that the primary event is a stepwise process and that the formation of P^+B^- takes only about 1 picosecond. This finding was subsequently confirmed experimentally.

Fortunately for us, a paper on the same subject [43] published in a meeting proceedings by Schulten, Karplus, Michel, Disenhofer and Huber turned out to be very problematic. Apparently, they had adopted our surface hopping approach, but did not figure out how to determine the energy gap between the two surfaces and how to evaluate the electrostatic energy in the protein (e.g., they did not place any water in and around the protein) and thus obtained an enormous error in the calculated energy and predicted a superexchange mechanism. Of course, this did not reduce the groundbreaking experimental findings by Michel, Deisenhofer and Huber, who received the 1988 Nobel Prize in Chemistry for their structural work.

The story of our parallel paths illustrates that it is difficult for the scientific community to establish which theoretical approach is correct, and with which theoretician one should collaborate. It also shows that the solution of complex biological problems may require knowledge of many unrelated approaches and, in particular, experience in what it takes to obtain reliable results. In our case, we were pleased to hear the theoretical chemist Michael Zerner say that we used all the methods known to mankind.

In 1988 I was invited to a Nobel symposium in Sweden that considered photosynthesis and other subjects, and was invited with the other participants to the Nobel award ceremony, where Laureates Michel, Deisenhofer and Huber were being honored. During the impressive Nobel banquet, I told someone at my table that I know something that might have slowed down one of the current Nobel prizes, which prompted someone else to chime in and ask who these Laureates were. At that point I realized that I shouldn't say anything that might be printed the next day in the Stockholm newspapers, so I said, "Never mind, it was just a bad joke." However, later in the evening, after congratulating Deisenhofer, I told him they might have encountered problems receiving the Nobel if the committee knew they had refused to give the structure to the scientific community (and no one could evaluate their findings and use it in a consistent way for understanding photosynthesis). I assumed that he realized that my main concern was about the problematic collaboration they had with Schulten and Karplus.

In later years, Bill Parson and I used our dispersed polaron model as well as a unique density matrix treatment to reproduce the observed kinetics at very low temperatures [44]. Here, again, we had to defend our research primacy from David Chandler of Berkeley, who got major credit as an expert in electron transfer by using, with Peter Wolynes, path integrals for estimating the approximated electronic coupling in electron transfer reactions. David appreciated the fact that I actually figured out the arguably more important issue, the microscopic basis of the Marcus parabolas as well as the ability to simulate the reorganization energy. But over time he became more competitive and tried to argue that I used just an approximated expression in my dispersed polaron treatment, while he presumably introduced the more rigorous so-called "spin Boson" treatment. This argument became popular in the chemical physics community, and I had to spend time proving that the dispersed polaron and the spin Boson treatments are identical [45] and that we happened to be the ones who figured out how to evaluate the quantum mechanical rate of electron transfer reactions in condensed phases, using classical trajectories. Unhelpfully, a reviewer of our spin boson paper wrote that everyone knows I'm right and I shouldn't kick dead horses which led the Nobel Laureate editor, Ahmed Zewail, to ask if I would withdraw the paper. I persuasively responded with a list of leading researchers who certainly knew the facts, but nevertheless, repeated Chandler's assertion.

In 1988 I was also invited to join a committee that reviewed the field of theoretical chemistry in Sweden, and had an interesting discussion with the head of the committee, Bengt Norden, who later was appointed chair of the Nobel committee in chemistry. Bengt was exploring the effect of electromagnetic radiation on human health, a challenging yet loosely defined task, but he also had many other interests. On a flight on a small airplane from Lund to Gutenberg, I told Bengt about my simulations of the dynamics of the first step in the visual process. Some years later, he claimed that my account of the femtosecond simulations contributed to his support for awarding the Nobel Prize to Ahmed Zewail for his work on femtochemistry.

In the 1980s, we lived in Culver City in a very nice gated community that was located around what was the old MGM lot. The burning of Atlanta in *Gone with the Wind* was filmed in the late 1930s near our neighborhood and some scenes in *The African Queen* were said to have been filmed in the pond beyond our condominium. There was a good neighborhood school for Merav and Yael, and in winter we traveled to Mammoth Mountain in the Sierra Nevada range to ski and toured the national parks in the summer.

Merav began her undergraduate study at UC Santa Cruz in 1988, and in 1992 Yael started at UC Berkeley. Neither showed any real interest in science, which was completely fine with me. But they did follow the academic life. Merav became a Professor in Health Sciences at Cal State Northridge, and married Raphael Efrat, a Professor of Accountancy who holds the Bookstein Chair in Taxation. They have two daughters, Maya and Danielle, and one boy, Neev. Yael is a Professor of Communication at Penn State and she is married to Idan Shalev, who is Professor of Biobehavioral Health. Tami took classes in art history and pottery, where she created many artistic works. She went on to get her teaching credentials at UCLA and began work in LA Hebrew schools, where she made a tremendous impact on her students. It was not surprising when we'd be walking in the mall and someone would suddenly stop Tami and say, "Mora (teacher) Tami, I really loved you as a teacher!"

XI

Moving Forward: 1990–2009

In 1990 I started writing a book on enzyme modeling, which initially followed some of the outline for the book I never finished with Mike, but ultimately expanded to become a textbook for graduate classes. It included small computer programs that demonstrated my approaches, moving from very basic quantum chemistry to modeling of molecules and then to modeling solution and enzymatic reactions with the Empirical Valence Bond (EVB) approach. The book concluded with a compressed chapter that examined (and mainly debunked) different proposals about the origin of enzyme catalysis. I was very pleased with the final product, *Computer Modeling of Chemical Reactions in Enzymes and Solutions*, which was published in 1991 [46]. In some respects, this book was far ahead of its time, although ironically a recent referee of a paper by a former co-worker asked him to remove a reference to it, claiming it was too old. Unfortunately, many did not get the key points of this work, which are as relevant now as they were in 1991.

Incidentally, Richard Friesner told me, when the book came out with the EVB programs (see below), that "now thousands of Indians will start using this method, and you'll be sorry." Unfortunately, that never happened.

Simulating G-proteins

In 1991 I learned about a major advance regarding the protein Ras p21, which is tightly involved in signal transduction (its mutations lead to defective activation that can cause many types of cancer). The signal

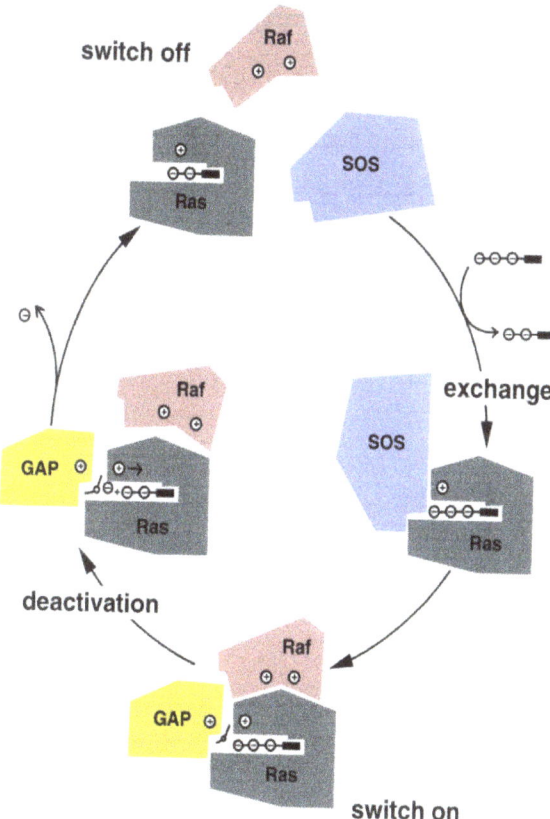

Figure 19. The cycle of activation of Ras p21. This protein serves as a switch, and in the on-state it binds GTP (designated by three negative charges). When the protein GAP binds Ras it drastically increases the rate of hydrolysis of GTP to GDP and inorganic phosphate, and the protein moves to the off-state. Mutations that reduce the catalysis by GAP leave the system in the on-state and lead to cancer. The figure is taken from [47].

transduction involves the hydrolysis of GTP (which serves as an activation switch) to GDP and inorganic phosphate (see Figure 19). A mutation that stops the GTP hydrolysis keeps the cell differentiating and thus leads to cancer. At the same time, I heard that a group at the Max Planck Institute at Heidelberg made significant progress solving the structure of this protein. So I traveled there from a meeting in Europe for a short visit, hoping to get the X-ray structure of Ras p21. I was aware of a fierce fight between the German group, Withinghofer, Pai and Goody, and Sung-Hou Kim from Berkeley, about the priority in the correct structural determination, but I had no intention of getting involved, having witnessed a similar fight between Klug's group with Alex Rich and Kim on the structure of tRNA. When I arrived, I got a nice presentation of the structure, but was told that they could not give me the coordinates. Again, as was the case with the bacterial reaction center, it turned out that they didn't have a clue who could do what. They gave the coordinates to Karplus' former postdoc Carmay Lim, who had moved to the University of Toronto, where Emil Pai was working. Of course, modeling Ras was a major challenge that required experience with actual phosphate hydrolysis, where we arguably had unique experience at that time (starting with the work with Johan Åqvist on staphylococcal nuclease). Since the last thing I wanted was to be scooped (even incorrectly) by Karplus or his associates, I contacted Kim, who was happy to provide me with his coordinates.

With Ralf Lange and Thomas Schweins, who came from Hannover, Germany, as M.Sc. students, we explored the nature of the catalytic mechanism of Ras p21. At first, we focused on the structural paper published by the German group, in which the proposed mechanism implied that an oncogenic residue, Gln 61, accepts a proton from a water molecule, in what can be called "Gln as a base mechanism". This proposal seemed to be problematic, since Gln is not a good base. However, to show that this is the wrong mechanism in the protein, we had to consider the basicity inside the protein. Fortunately, our EVB method provided a powerful way to explore this issue, with rather straightforward, yet reliable, calculations. That is, we used the EVB to show that in the enzyme the barrier for the proton transfer from the catalytic water molecule to Gln 61 (to form the nucleophilic OH^-) is significantly higher than the observed barrier for the actual enzymatic reaction. This meant that the reaction cannot involve a proton transfer to Gln. Our resulting article was published in 1992, through accelerated publication, in *Biochemistry* [48]. Interestingly, a reviewer went to great lengths to contend that we took Kim's side in the structural debate, and intentionally published a mechanism disputing the one proposed by the Hidelberg group. I responded that I do not consult with crystallographers in my mechanistic simulations. At any rate, our mechanism was supported by experiments with an unnatural amino acid replacing Gln 61.

In subsequent years we kept advancing our modeling of phosphate hydrolysis reactions and in particularly GTP hydrolysis (e.g., [49]). This led to studies of other GTPase and ATPase proteins that play key roles in molecular biology, and to gradual progress in describing the energetics of phosphate hydrolysis reactions in solution (using *ab initio* studies) and proteins (using the EVB to transfer the solution surface to the protein) [50]. We extended our studies to the activation of the ribosome by the elongation factor, EF-Tu, and the EFG protein. Lynn Kamerlin (my post-doctoral associate at that time) and I summarized our work in a major review "Why Nature Really Chose Phosphate" [47].

Overall, we arguably led to a paradigm shift, establishing that studies of phosphate hydrolysis and phosphoryl transfer, which had been the subject of countless experimental works, must be analyzed by computational studies. These cannot substitute for experimental studies, but they are crucial in the interpretation stage. This point has been established by our studies of linear free energy relationships and analysis of other proposals that were postulated by the experimental community and found to be problematic.

Bacteriorhodopsin and Rhodopsin

Bacteriorhodopsin is a membrane protein that pumps protons across the membrane after absorption of light, when it uses light energy for photoisomerization of a bound Schiff base of retinal. The photoisomerization changes the electrostatic potential in the protein and leads to proton pumping. This system provides a key model for the energetics of proton pumping and for light-induced photoisomerization. It also offers crucial information on the related action of rhodopsin.

I began studying bacteriorhodopsin in 1979 [51], exploring the energetics of the pumping process and elucidating key principles of light-induced charge separation. This clarified for me key principles of bioenergetics, including the role of pK_a changes and their coupling to the overall free energy storage. However, although these studies made important predictions, they were conducted before the structure was known. When the atomic structure emerged in 1990 I used this structure in clarifying some aspects of my 1976 vision paper, taking the newly elucidated structure and propagating excited-state trajectories. I performed the simulations in order to establish that the initial light energy dissipates very fast and thus does not influence the quantum yield. This was an important study since researchers who challenged my 1976 vision paper argued that my finding was contradicted by "experimental facts". In particular, they claimed that while the observed quantum yield is independent of the excitation wavelength, "everyone knows" that with more energy we will yield different results. This widely quoted presumption reflected the superficial conclusion that would be reached by looking at a one-dimensional trajectory that behaves in an inertial way, whereas if we start with larger energy we reach the minimum with faster motion, just as a ball would roll on a downhill slope. Already in 1975–1976 [22], I used multidimensional trajectories where the energy was found to dissipate very fast in the chromophore itself. Obviously, the dissipation should be larger when the protein is included, and indeed we were able to show that in a realistic bacteriorhodopsin active site the chromophore moved downhill on the excited-state surface in less than 100 fs, but still dissipated all of its kinetic energy [52]. Thus, we get a similar quantum yield for different initial energies.

In another study, I used the bacteriorhodopsin simulations to show that the approach I used in 1976 for evaluating the surface crossing in rhodopsin captured the conical intersection effect that became popular in later years. That is, studies that emphasized the crucial importance of conical intersections in photochemistry suggested that all the excited state trajectories in rhodopsins lead to a perfect transfer to the ground state through a conical intersection funnel. I felt, however, that in a rapidly fluctuating protein site, it is unlikely that all the trajectories would run through a perfect funnel. Therefore we propagated many QCFF/PI trajectories (a task that could not be performed with *ab initio* approaches) and "proved" that it is impossible for all trajectories to pass through an exact conical intersection [53]. Our point was based on logical considerations rather than on exact calculation, which could not be accomplished in the foreseeable future. Since the energy gap between the two relevant surfaces cannot be determined exactly inside the protein, we shifted the energy gap down and up until we obtained maximum crossing probability. At that point, some of the trajectories crossed before the ground state surface reached its maximum value and were deflected and reversed the direction of rotation, which thus led to a reduction in the quantum yield. When we raised the energy gap we obtained lower quantum yield, since we moved the system from a point of maximum crossing probability.

We also performed studies of the effect of the photoisomerization of bacteriorhodopsin on the proton pumping process. These confirmed my prediction from 1979 [51] that the movement of the Schiff base during the photoisomerization decreases the electrostatic stabilization of the ion pair between the Schiff base and the nearby ionized aspartic acid (and changes the pK_a of both the aspartic acid and the Schiff base), whereas the actual isomerization and the corresponding presumed change in conjugation of the chromophores leads to a rather small change in the pK_a of the Schiff base.

How do Enzymes Really Work

After having aimlessly chosen to study chemistry, as I've described, my first clear interest in entering the world of science was to discover the origin of the catalytic power of enzymes. The puzzle of how enzymes catalyze chemical reactions by many orders of magnitude has been around since the early 20th century. It has been customary, and convenient, to attribute the solution to Linus Pauling, who allegedly suggested that enzymes stabilize transition states [54]. However, his insightful suggestion amounted to saying correctly that enzymes reduce the activation barrier, without being able to ascertain how. Pauling actually focused on van der Waals forces, which amounted to destabilizing the ground state rather than stabilizing the transition state. Of course, he did not have the structure of any enzyme and could not explore his proposal. In fact, although many cited the idea of transition state stabilization, almost all clear catalytic proposals amounted to ground state destabilization. And so the puzzle of the origin of enzyme catalysis remained one of the most important fundamental problems in biology. Realizing that the elucidation of the factors that determine the catalytic power of enzymes must be addressed by some type of structure energy modeling, we developed the QM/MM approach and established qualitatively the importance of electrostatic stabilization in lysozyme. But a quantitative validation was still needed. At that point I realized that the physical organic community had not used a clear energy-based definition, making it hard to quantify the problem. It even appears that in the early 70s, most researchers did not believe in transition state theory as a way to obtain the activation free energy from the rate constant. This was due perhaps to the belief that the transmission factor (the coefficient in the transition state theory expression for the rate constant) is strongly dependent on the orientation of the reacting fragments, as is the case in gas phase reactions. However, in condensed phases, the transmission factor is almost constant once the barrier is higher than a few kcal/mol. Furthermore, in many cases we did not have rate constants or energy estimates for individual steps in solution, and obtaining the relevant energies required energy estimates and quantum mechanical calculations. Another critical issue was the reliability of the molecular orbital treatment used in our QM/MM approach.

To resolve this critical problem, in 1980 I introduced, with the help of Bob Weiss, the powerful EVB method [27], based on the view that it would be hard to get absolute activation barriers for enzymes by any standard QM/MM approach. We represented the reaction by a simple Hamiltonian, with the physics of bond breaking and bond making, and with a consistent coupling to the surrounding solvent, insisting on calibrating the relevant Hamiltonian parameters on reference reactions in solutions. In this way, moving the calibrated EVB system from solution to enzymes converted studies of the enzymatic reaction to studies of *differences* in the free energy of the reacting system in the enzyme and in the given reference solution environment, rather than attempting to evaluate absolute activation barriers.

The original EVB formulation used the protein dipoles Langevin dipoles (PDLD) model with some *ad hoc* assumptions about the self-consistent relationship between the solvent and solute polarization [27]. Later, in the mid-80s, we moved to an all-atom molecular dynamics model, with a fully consistent free energy perturbation approach that took into account non-equilibrium solvation effects [55], long before the treatments of such effects by other researchers moved from theoretical discussions to approximated treatments. Considering the limited computer power of that time, we have to judge the EVB convergence carefully by the accumulation of encouraging results.

At some stage, due in large part to Johan Åqvist, we started to realize that the EVB results could actually converge to stable and reliable results. Thus, the EVB appeared to be a powerful and reliable approach that allowed us, and others, to explore many enzymatic reactions, mutational effects and enzyme design. Many researchers had trouble believing we could get reasonable results for enzyme catalysis. For example, at the start of a 1992 Faraday discussion in Cambridge that focused on enzyme modeling, the organizer, Professor David Buckingham, declared, "Of course we cannot even model the H_2 molecule so how can we model enzyme action." However, the distinction is that the EVB does not model the enzyme by a first-principles approach, but evaluates the difference between a reference reaction and the enzymatic reaction (which is mainly due to electrostatic effects). This made the EVB an extremely robust approach. In fact, the spirit of the EVB was close to the idea of an asymptotic wave function that I promised my friend in physics class in 1965. When I received the Nobel Prize, I was told half-jokingly by the committee member in charge of quantum chemistry, that "We didn't give you the Prize for your EVB."

At any rate, one of the EVB's most important additions was the ability to quantify the main contributions to enzyme catalysis. More specifically, our studies of different enzymes indicated that in all cases the primary contribution comes from the difference between the electrostatic stabilization of the transition state in the enzyme and in solution. The difficulty was to understand the origin of this effect. Here we exploited the ability of the EVB to evaluate Marcus' reorganization energy (the energy released when the environment relaxes from its orientation in the reactant state to its configuration in the product state) and compared the reorganization energy for the reaction in the enzyme and in solution. We found that the reorganization energy is significantly larger in solution than in the enzyme active site

[56]. Since larger reorganization energy leads to larger activation barriers (as long as the reorganization energy is larger than the reaction energy), we have a clear explanation for the catalytic effect. This issue becomes somewhat complicated by the finding that the interaction energy between the environment and the charges of the substrate in the transition state are similar in enzyme and solution. Thus, the reason for catalysis is not associated with stronger interaction between the enzyme and the substrate's transition state, but rather with the fact that in water the solvent dipoles have to rearrange significantly to stabilize the transition state charges, while the enzyme has its polar dipoles already preorganized toward the transition state charges and so pays less in stabilizing the transition state. The preorganization origin of enzyme catalysis is illustrated in Figure 20.

It should be noted that despite the current popularity of the electrostatic idea, for a long time it was not considered a valid proposal by the physical organic chemistry community. Of course, we could also note that my undergraduate work in the Technion concluded incorrectly that electrostatic effects do not help catalysis. The main argument against the importance of electrostatic effects came from careful experiments that explored the effect of covalently bound charged groups on chemical reactions (see [56]). Such studies (e.g., Tom Bruice's works) found very little electrostatic effect and thus excluded the idea of major electrostatic

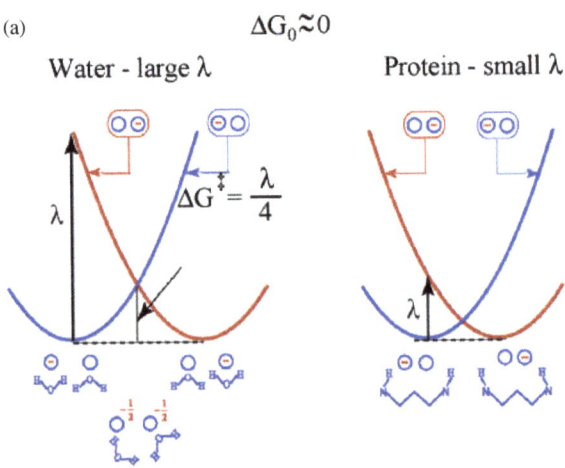

Figure 20.   (*Continued*)

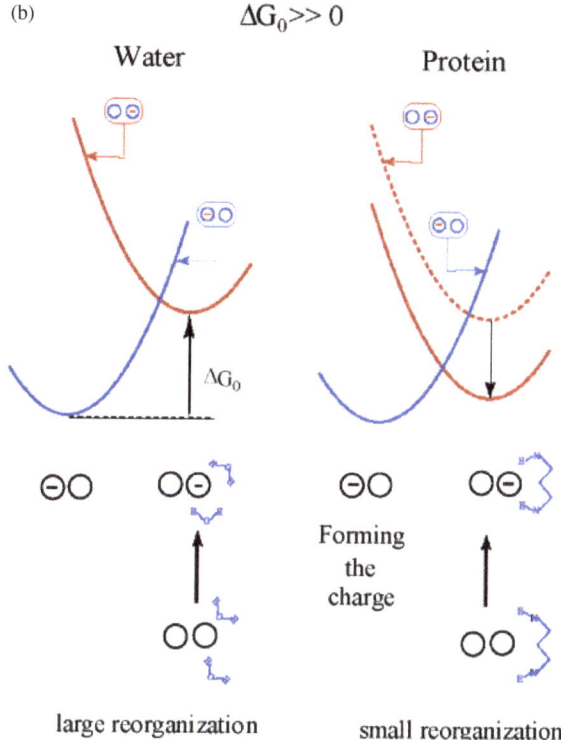

Figure 20. Explaining the role of protein preorganization in enzyme catalysis. (a) The figure describes the free energy along the reaction coordinate (rotation of solvent molecules or rearrangement of the enzyme groups) for the reactant state (– 0) and the product state (0 –). The intersection of the parabolas represents the reaction barrier. The barrier, for the case of equal energy of the reactant and product, is given as ¼ of the reorganization energy. Now, since the enzyme dipoles (right) reorganize less than the solvent dipoles (left) we have smaller reorganization energy and smaller barrier in the protein. (b) The same situation when the energy of the product is higher than that of the reactant. Here again the protein has a smaller barrier. The catalysis is due to the reduction of the reorganization energy in the folded protein, which reduces the barrier relative to the situation in the bulk water. The figure is taken from [56].

catalysis. What was overlooked in such analyses was the realization that the dielectric "constant" can be very different in an enzyme active site than in solution.

In addition to using computer simulations as a guide for determining the most important factors in enzyme catalysis, we used them to eliminate previous proposals that were supposed to account for the power of enzymes. Some of these systematic eliminations are considered below.

The Circe Effect Does Not Account for Enzyme Catalysis

In my effort to understand and clarify enzyme catalysis, and to establish my findings, I became involved in several "controversies". In each case, I used computer simulation to examine proposals that were accepted by the biochemical and chemical community and described in major textbooks. But none of those proposals were defined consistently by energy-based considerations and thus could not be proved or disproved. One instructive example was the search for the origin of the catalytic power of orotidine decarboxylase (ODCase), identified by Dick Wolfenden as one of the most effective enzymes, with an enormous catalytic effect of more than 20 kcal/mol. It includes a substrate with a negatively charged carboxylate group that is converted to CO_2 while the substrate ring becomes negatively charged. The original proposal for the action of this enzyme invoked the classical desolvation idea, suggesting that the enzyme active site is like an oil drop that destabilizes the charged ground state and reduces the activation barrier. After this proposal was published in *Science*, Jan Florian and I published a *PNAS* paper [57] repeating our point that enzymes are not oil drops, and predicting that when the enzyme structure is elucidated it will be found to have a very polar (rather than oily) active site.

As we forecasted, when the structure was solved it was found to be very polar, with a protein aspartic (Asp) group pointing toward the substrate's negatively charged carboxylate. At this stage another *Science* article appeared, postulating again that the catalysis is due to ground state destabilization (GSD), but now because of the repulsion between the ionized asp residue of the protein and the substrate's carboxylate. The weekly journal *Chemical & Engineering News* published a story about this proposal, including a great endorsement from a senior biochemist, who said that he would start to teach about this enzyme in his class as a key example of GSD. I was quoted, however, stating the simple fact that if the ionized Asp were brought near the substrate's carboxylate the repulsion between them would force the Asp to accept a proton (and thus lose all the GSD). Soon after that paper, we performed careful calculations of the reaction in the enzyme and solution, taking into account the fact that the reaction in the enzyme includes a positively charged lysine (Lys) residue that participates in the actual reaction. By including the Lys in the reacting system, we could account for the catalysis with a very large transition state stabilization (TSS) [58]. The story became even more interesting when the catalysis by this enzyme continued to be cited as an example of the famous "Circe proposal" of the great biochemist Bill Jencks, where enzymes work by GSD due to the binding energy of distant groups of the substrate that are not parts of the chemical region [59]. These groups are assumed to pull the unstable part of the substrate into the active site (in the same way that Greek mythology's Circe lured Odysseus' men into her mansion, drugged them, and transformed them into pigs). The Circe proposal has been very popular and used as a general explanation for the catalytic power of enzymes, without providing clear and correct thermodynamic considerations.

We argued, based on energy calculations, that the non-chemical part (a "tail" chain with a negatively charged phosphate at its end) cannot account for the destabilization needed for catalysis [60], but this did not convince many people. However, John Richard and his co-workers, who were perhaps originally motivated to show we were wrong, performed a great relevant experiment that supported our point of view. They chopped the substrate's tail and indeed found that the catalysis disappeared, which seemed to be consistent with the GSD Circe idea, since without the presumed distanced strain the substrate should lose the destabilization of the chemical part and the catalytic effect. However, they then did an extra experiment and added a negatively charged inorganic phosphate instead of the tail, and then suddenly the enzyme started to catalyze the reaction [61]. This demonstrated that the bond to the phosphate (and the corresponding presumed distanced strain) is not the reason for the catalysis. In other words, enzymes do not work by using distanced binding energy (Figure 21).

1. Using Binding Energy for Catalysis

Unstable ground
state and thus less
effort to go to the TS

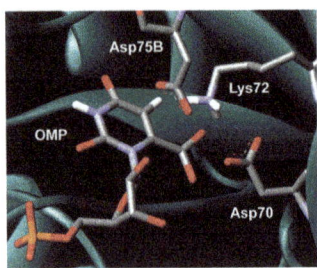

**The PO$_3^{2-}$
Binding Site**

2. No Catalysis

No strain no ground state
destabilization and thus no catalysis

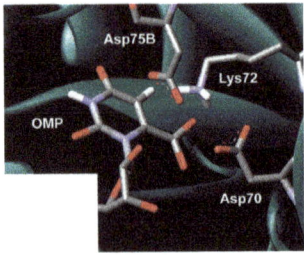

Figure 21.   (*Continued*)

3. Good Catalysis

However, without any strain
we get catalysis

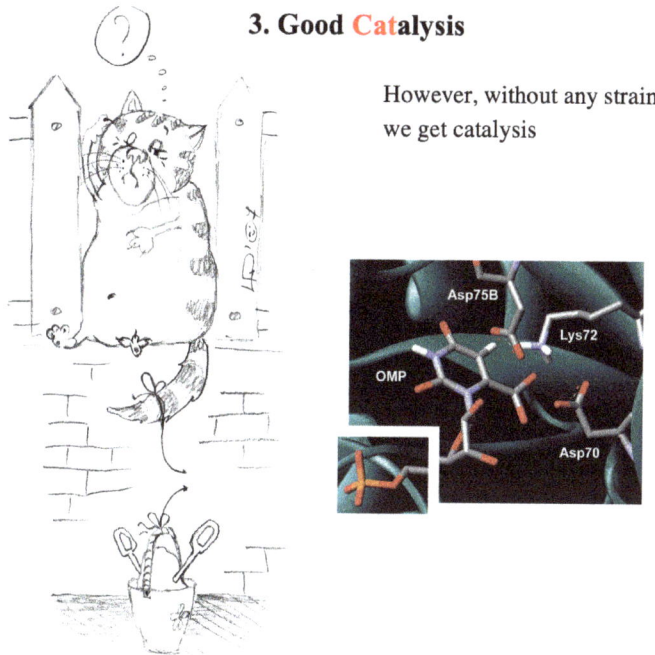

Figure 21. A description of the experiment that actually disproved the Circe effect. (1) Description of the ODCase enzyme substrate system, an extremely powerful catalyst, for which the Circe proposal assumes that the binding of the phosphate tail pushes the substrate to a state with a very large GSD (due to the presumed repulsion between the substrate carboxylate and Asp 70). (2) After chopping off the tail group the catalysis stops, which seems fully consistent with the Circe proposal since the presumed GSD disappears when the tail does not pull the substrate to the position where it has a large GSD. (3) Surprisingly, adding inorganic phosphate restores the catalysis, meaning that the catalysis is not related to the strain between the tail and the reactive part. The figure was drawn by Mrs. Psiliakov.

Proton Tunneling in Enzymes Does Not Contribute to Catalysis

At various stages of my scientific studies, I had to address the criticism that my research is problematic since it does not consider various small effects. A good example involved my anticipation, in 1990, that sooner or later other scientists will be able to calculate quantum mechanical tunneling corrections for proton (or hydrogen) transfer in enzymes. At this point some researchers may claim that my findings should be ignored, since I did not consider "a key effect". But I was quite sure that the quantum mechanical tunneling corrections are similar in the reference solution reaction and in enzymes (and thus are not likely to contribute to catalysis), thus I concluded that my own calculations must establish this quickly. I started, with the help of JK Hwang, to look for approaches for modeling tunneling. We already had a method for the cases of small coupling between the product and reactant states (namely, the dispersed polaron approach I used in studying electron transfer reactions), though here we had a strong coupling case, which is much more challenging. After several attempts, we adopted the Feynman path integral model, using Michael Gillan's great idea [62] that the probability of the centroid of the path integral quasiparticles (the particles that replace each atom) should provide the profile of the quantum free energy. We then combined this proposal with a very powerful trick of using the classical trajectories of the actual atoms, rather than those of the quasiparticles. The actual results of the path integral distribution were then obtained by perturbation from the classical path to the quasiparticles path [63, 64]. With this enormous time saving we managed to be the first to simulate quantized proton transfer in enzymes and solutions, which allowed us to explore the role of tunneling in enzyme catalysis (see below).

Although my quantum classical path approach was adopted in different incarnations by various researchers, it also led to a less welcoming response. For example, a reviewer of our 1996 *JACS* paper [65] on the quantum corrections in the enzyme carbonic anhydrase wrote that it is well known that I always lie about my findings, since no one can model enzymes. He went on to assert that my fake science can be clearly established, since I deny the existence of quantum mechanics (perhaps missing the point of using classical trajectories as a reference for quantum calculations). I responded that this would be his best chance to finally expose me and show that I have been falsifying my results. Since I also published a one-dimensional verification (where I actually reproduced the quantum results at low temperature), he could clearly check my calculations, and while he wouldn't model any multi-dimensional systems, he should be able to expose me by rerunning my approach on the very simple one-dimensional model. The editor promised to send him my response, but I doubt it reduced the reviewer's hostility.

The ability to simulate quantum mechanical corrections for proton and hydrogen transfer reactions became very useful in my effort to show that such effects do not contribute to enzyme catalysis. That is, in view of the difficulty for many to accept my arguments about the preorganization effects, there has been a strong tendency to accept exotic notions, such as that nuclear tunneling is a major factor in catalysis. This idea was originally proposed with the model that considered the chemical barrier as a narrow rectangular potential that should become narrower in the enzyme due to the steric effect of the active site and increase the tunneling contribution. This concept was presented in high-impact journals, in part due to research that found large tunneling corrections in some enzymatic reactions. However, we demonstrated that the tunneling effects that exist in the enzyme also exist in the same reactions in the reference systems in water without the enzyme, and thus have no catalytic effect. Furthermore, we showed that enzymes cannot apply large steric effects to compress barriers. Despite our papers that quantitatively reproduced the isotope effects in enzymes and solutions, the tunneling idea continued to be popular. This became particularly problematic when we found that in proton transfer reactions the barrier actually goes down upon compression rather than narrowing, and that the tunneling increases when the distance between the donor and acceptor increases rather than decreases [66, 67]. Our finding should have put to rest the catalytic tunneling idea, but its proponents chose to ignore our clear physical and conceptual findings and never mention them.

Enzyme Dynamics and Catalysis

In 1978 I attended a physics seminar at Caltech. The speaker was Prof. Giorgio Careri, an Italian physicist who promoted quite early the importance of dynamics in enzyme catalysis. I sat in the front row near star physicists Richard Feynman and Murray Gell-Mann (whose faces I didn't recognize at the time) and Careri started to talk about magnetic properties of enzyme powder, and then ventured to suggest that because of these features, dynamics must be important in catalysis. This sounded interesting to Feynman and Gell-Mann, who probably had no clue about enzymes. I whispered loudly that this is "nonsense", since I could not believe Careri's logic — there was no relationship between the observation and the proposal — which Harry Gray, who attended the lecture but did not know me back then, subsequently mentioned at several events.

The idea that enzymes work by dynamics [68–72] was easy to spread since it seems to offer an explanation of an effect that is hard to understand. As Henry Kissinger has said, some people try to win an argument by suggesting an explanation so complex that neither side would understand it. And so for over a decade the concept of special dynamical effects was the most prominent idea about enzyme catalysis, including the suggestion that dynamics is the key to enzymology in the 21st century, and was published repeatedly in all the high-profile journals. Believing this proposal is extremely problematic — it was poorly defined and lacked any critical analysis — I spent a significant effort trying to explain how misguided it was, starting in 1984 [73], and then particularly from 2000–2010 [38, 74, 75]. I began by trying to clarify what the dynamical proposal must mean, without making an ill-defined circular argument. Of course, we all know that atoms are moving, but for dynamics to help in catalysis it must have a larger effect in the enzyme than in the corresponding reference reaction in solution. The usual definition of a dynamical effect is the classical correction of transition state theory due to motion back and forth on the top of the barrier. However, such corrections were found to be insignificant and of the same magnitude in water and in the enzymes. Apparently, the only way to define what the dynamical idea must mean is with the notion that in the enzyme active site the reactive motions are inertial. That is, the motions from the ground state to the top of the barrier are such that the trajectories of the binding process retain their kinetic energy and are used to cross the reaction barrier. Fortunately, we were able to show [74] (by using a lower dimensional model with a renormalized friction) that enzymatic reactions with barriers that are higher than a few kcal/mol cannot have any appreciable dynamical effects. Once the barrier is sufficiently high, the trajectories that lead to the reactive complex move first back and forth on the bottom of the chemical barrier, losing any inertial memory of where they came from and what their kinetic energy was. They must wait for the thermal chance (the Boltzmann probability) to cross the barrier. Apparently, enzymatic reactions follow the Boltzmann probability (see Figure 22) and cannot use dynamical effects to help in their catalytic power.

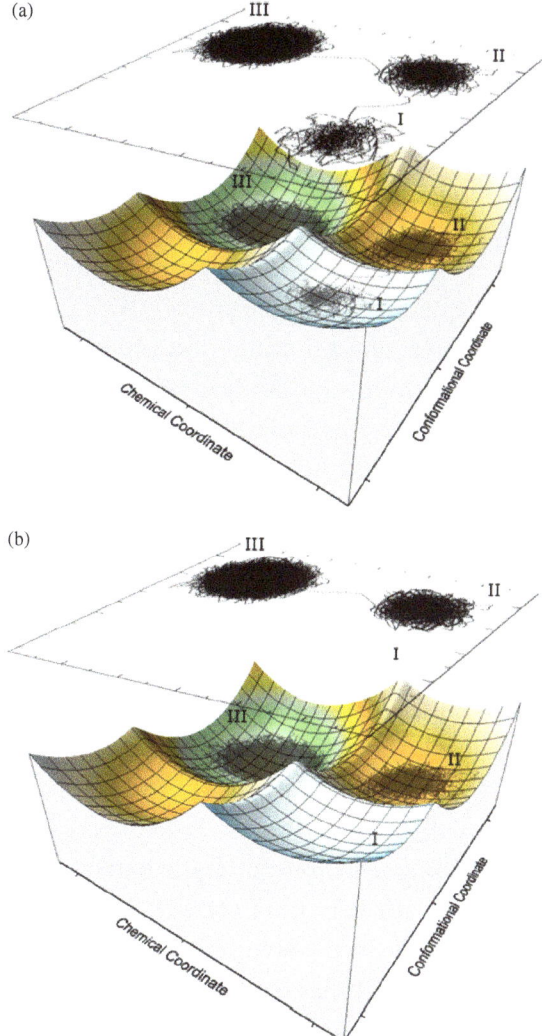

Figure 22. Illustration of the fact that dynamics cannot help in enzyme catalysis. (a) shows a trajectory that starts at the high-energy, unbound enzyme-substrate state and continues to the bound substrate (ES) state with a large kinetic energy. However, this trajectory stops before the chemical barrier and moves back and forth for a long time, losing all the kinetic energy. Eventually the thermal energy generates a reactive trajectory that passes the chemical barrier according to the Boltzmann distribution. (b) The trajectory starts from the ES state, when again the thermal energy leads to a reactive trajectory of the same kind observed in (a). Comparing the two figures establishes that the excess kinetic energy in (a) does not help catalysis (since we get the same result in (b) without this kinetic energy). Thus, dynamics does not help catalysis. The figure is taken from [74].

It is Hard to Explain Enzyme Catalysis Without Consistent Energy Considerations

One of the problems that slowed down the building of consensus about enzyme catalysis was probably a gap in rigorous training in energy considerations. While it might be hard to accept, it seems that although most researchers know how to draw diagrams with energy barriers, many are not fully familiar with the corresponding energy-based concepts (and much less with their quantitative implications). For example, I found out that in the case of the reaction of serine proteases, some biochemists and some textbooks assume proton transfer from serine to histidine is a downhill step in the enzyme (and thus looks like a step that pushes the reaction forward). In fact it is an uphill step both in the enzyme and in solution. The difficulty appears to be in realizing that catalysis involves having to spend less thermal energy to *climb* the barrier in the enzyme than in the same reaction in water. I confronted this rather common problem many times when I tried to explain my findings to reporters. Suggesting that catalysis reflects steric strain (like the effect of pushing a door supported by springs) is easily understood (even as it is an incorrect explanation for enzymes), whereas the argument that electrostatic effects reduce the barrier and thus lead to catalysis appeared to be much harder to grasp intuitively. In view of these difficulties, I would sometimes present explanations that look intuitively correct in my undergraduate classes and only then told them that the real story is more complex and requires careful free energy considerations. One of my favorite examples was the discussion of Jencks' popular proposal [76] that enzyme catalysis is due in large part to entropic contributions. Jencks suggests that when the substrate is in its ground state in water it has a large entropic contribution (the reactive groups are free to move) and thus negative free energy. However, when the system goes to the transition state in water the motion becomes very restricted and the entropy contribution becomes small. On the other hand, in the enzyme the motion is already restricted in the ground state and thus we do not have a large entropic contribution in either the ground or the transition state. This sounds like a very logical example of ground state destabilization and is easy to understand — but it is incorrect, since it misses a key point. The restriction in motion upon going from the ground state to the transition state in water is much less than what it was thought to be, since only one degree of freedom is actually frozen [77]. Thus the transition state in water also involves major entropic contributions. Furthermore, the entropic proposal actually applied to the binding entropy and not to the catalytic effect. It is instructive to note that Page and Jencks' article appeared in *PNAS* in 1971 [76], gaining wide approval. I remember that during my time in the MRC, leading scientists would argue about it at tea break, questioning the idea that the entropy of a substrate in water is similar to the corresponding entropy in the gas phase (a very reasonable assumption). None of them, however, examined the relevant free energy diagram, which did not include the transition state entropy, and could not realize that the proposal was based on an incorrect analysis [38].

The Impact of Experimental Support for the Electrostatic Ideas

An experimental verification of theoretical predictions is often very useful. In my case I had several such breaks, one of which had its genesis at a meeting of the American Chemical Society around 2008, during a conversation with Steve Boxer from Stanford, whom I knew well from the photosynthesis field. Boxer had started to study enzymes using vibrational Stark effect measurements and was collaborating with Dan Herschlag, one of my most hostile competitors. "In about one year you will either love me or hate me," he said to me. To which I responded, "There is no way you'll find anything that would consistently contradict my electrostatic concepts, so I don't think I will have to hate you." Over the next few years Boxer found more and more detailed experimental proofs that the catalytic effect of keto-steroid isomerase (one of Herschlag's and my fiercest battlegrounds) is almost completely correlated with electrostatic effects and consistent with our studies. His findings represented significant support for the idea that enzyme catalysis is due to electrostatic effects.

I would often note that this reminds me of Newton and Halley. As the story goes, no one really read Newton's *Principia* until the calculus prediction of the path of Halley's Comet was confirmed by Halley. As this anecdote indicates, it's difficult to gain acceptance for concepts that emerge from theoretical considerations that were not presented as confirmed predictions. As a great believer in computational predictions, I encountered this problem repeatedly. A good example was our publication of the first paper that dealt with the mechanism of DNA polymerase. When the structure of this important enzyme started to appear, I managed to get the X-ray structure of the exonuclease domain in the Klenow fragment of *E. coli* DNA polymerase I from Lorena Besse (who worked with Tom Steitz), and to present the first consistent theoretical study of the two-metal mechanism. But *JACS* rejected my paper based on the argument that I had not made any prediction of mutations. I believed this was unjustified, considering this was the first consistent analysis of the reaction of DNA polymerase, as well as systematic proof that the two Mg^{2+} ions do not work by some steric strain but by electrostatic effects. I called the chief editor, Alan Bard, and asked to replace the editor working on my article, but Bard repeated the argument that I had not made any predictions, but my response got him thinking: "I bought your book on redox potentials, which not so many others probably read very carefully," I began. "Let's say I have a computer program that can reproduce the observed redox values you report, regardless of who runs the program. Would you still demand that the program make predictions?" Bard hesitated, and then conceded: "Well I guess you're a special case," he told me, and then sent the paper to another editor, who accepted it [78].

The EVB Has Many Presumed Fathers

Our EVB method was very successful and started to be used by many groups. Unfortunately, as frequently happens, some groups changed its name (AVB, MS-EVB, KH Theory, and more), trying to create the impression that they were introducing something new. Our method was even once attributed to an obscure researcher who had nothing to do with its development. In a similar, more serious case, the monumental electron dot model of Gilbert N. Lewis was called the Langmuir model by the physicists who didn't like Lewis. With this in mind, Florian wrote a 2002 paper [79] that classified all the "new" methods by their names and their origin. And still the problem persisted.

In 2008, Don Truhlar, Jiali Gao and their coworkers tried to drastically minimize our credit, claiming [80a] that it was impossible to reproduce our results even for the gas-phase energetics of the reacting part of the enzyme dehalogenase. But instead of asking us for the computer program, they tried to rewrite the corresponding code, and unfortunately used an incorrect sign in one of the key terms. Their long paper, which may have looked relatively innocent to a superficial reader and was accepted without showing it to us, presented a series of misleading accusations. In just one example, they argued that the EVB should have been calibrated on *ab initio* results, knowing very well that we had been performing such a calibration since 1988. This was a major challenge for us, since it clearly implied that all of our enzyme calculations were incorrect. It was particularly problematic for me since such an insinuation could jeopardize my candidacy at the time to the USA National Academy of Science. Thus, we were forced to decisively respond. Fortunately, George Schatz, the chief editor of the *Journal of Physical Chemistry*, agreed to carefully read Truhlar's paper and realized what was actually written there, despite its veneer of polite discussion. He allowed us to write a long, point-by-point response letter, titled "On Unjustifiably Misrepresenting the EVB Approach While Simultaneously Adopting It" [80b], which left no room for ambiguity, including showing that with the correct sign the correct results were quite nicely reproduced. Truhlar tried very hard to stop the letter's publication, claiming it was criminal to have a computer program that could not be replicated. We responded that no modern program, with its many parameters and features, could be reproduced without the original code and output for comparative validation (both of which were available upon request in our case). Basically, you should never claim that someone's method and code are incorrect without making sure that you are using the same method. Otherwise it is a very serious slander that has no place in science. We also pointed out that one might suspect that the sign error in reproducing our results was intentional (since we were not contacted), which is a clear ethical problem. Instructively, our vigorous response was effective in blocking the attempts to discredit the EVB, which remains a widely used and extremely effective tool in modeling enzymatic reactions. It also turned out to be a powerful means to understand the general trend in chemical reactivity.

The National Academy of Science

In the early morning in late April 2009, I got a phone call informing me that I was elected to the USA National Academy of Science (NAS). Although I knew that I was on the NAS shortlist, it wasn't certain. In fact, I always thought that in view of the controversy regarding my research, I would have to be awarded the Nobel Prize before being elected to the NAS. The NAS vote is much more political than the Nobel, and being at a university like USC, with very few NAS members to support you, made it extremely challenging. I was elated and told Tami it was a miracle. In fact, some of my friends were amazed that someone with my political problems could be elected, one even asking whether I was surprised that I was elected while one of my enemies was still alive. I responded I was surprised that I was elected while I was still alive. A huge advantage for me was publishing in the *PNAS*, which finally reduced the need to deal with the frequent hostile and irrelevant reviewer comments.

At this stage, I became significantly more distinguished at USC, where I was allowed to have only half the standard teaching load. I also became more relaxed, less worried about my scientific struggles.

XII

Multiscale Modeling of Large Biological Systems

Many key biological systems involve large protein complexes and/or membrane protein systems. In such cases simulations by all-atom models present a major challenge, in terms of the computer resources needed for proper convergence. Our original 1975 coarse-grained (CG) simplified folding model [17] did not include the proper electrostatic energy terms so crucial for determining protein functions. So we refined it, and added crucial electrostatic energy terms calibrated on relevant experimental results, such as protein stability, solvation energies and the energy of amino acids insertion into membranes [81, 82].

The refined CG model allowed us to explore the action of major biological systems. This included simulating molecular motors, such as ATPase, where we used it to determine the origin of the unidirectionality of the energy conversion (see below). We also explored the action of voltage-activated ion channels that play a major role in many key body functions, such as brain activity. In addition, we simulated membrane protein action, including the ribosome-assisted insertion of proteins through the membrane (via the translocon) [83].

Biological Motors and the Conversion of Chemical Energy to Mechanical Energy

Biological systems use nanomotors, which orchestrate many cellular actions. Such motors use chemical energy from the conversion of ATP to ADP plus inorganic phosphate to perform mechanical work. Interestingly, this process involves the use of pH gradients. In some respects, understanding biological motors has been motivated by the interest in generating man-made nanomotors that would allow one to control, for example, the science-fiction task of surgery in the cellular space. Yet biological energy conversion is a more fundamental issue.

The energy conversion in molecular motors started to interest me back at a Johnson Foundation meeting in 1978, when I met Peter Mitchell and tried to tell him over breakfast about my electrostatic concepts. It was just before he received the Nobel Prize, and I learned about his ideas on the key role of the electromotive force of the pH gradient. Subsequently, I tried to obtain more solid information about the energetics involved and the corresponding chemical barriers, which I unfortunately discovered was largely missing from the available discussion and analysis. Although I generated and maintained a dedicated file for this project, I did not progress significantly until 2002, when I got involved in analyzing the energetics of ATPase, armed with the data I gained from studies of phosphate hydrolysis in solutions and GTPases.

Between 2003 and 2011 I collected and classified experimental and theoretical information on the energetics of the different steps in F1-ATPase, and decided to take on the analysis of the overall conversion of chemical energy to mechanical energy (Figure 23). From my perspective, this could not be clearly analyzed without a well-defined structure/energy description that was not available at that time. Our main progress began with the work of my postdoctoral associate Shayantani Mukherjee and me [84], in which we used our CG model and reproduced the so-called "catalytic dwell", where the rotation of the central stalk of the motor stops and the chemical reaction occurs. In subsequent works, we quantified the conversion of the chemical energy to torque and studied the origin of other features of the action of the motor [85]. Our research presented the first realistic elucidation of the overall free energy landscape for biological motors. We also demonstrated that the motor's action is determined by this landscape rather than by some random Brownian motions, as many researchers had assumed. Furthermore, we succeeded in modeling the conversion of the pH gradient to ATP synthesis by the action of the F_0 part of the ATPase/membrane system [86]. In some

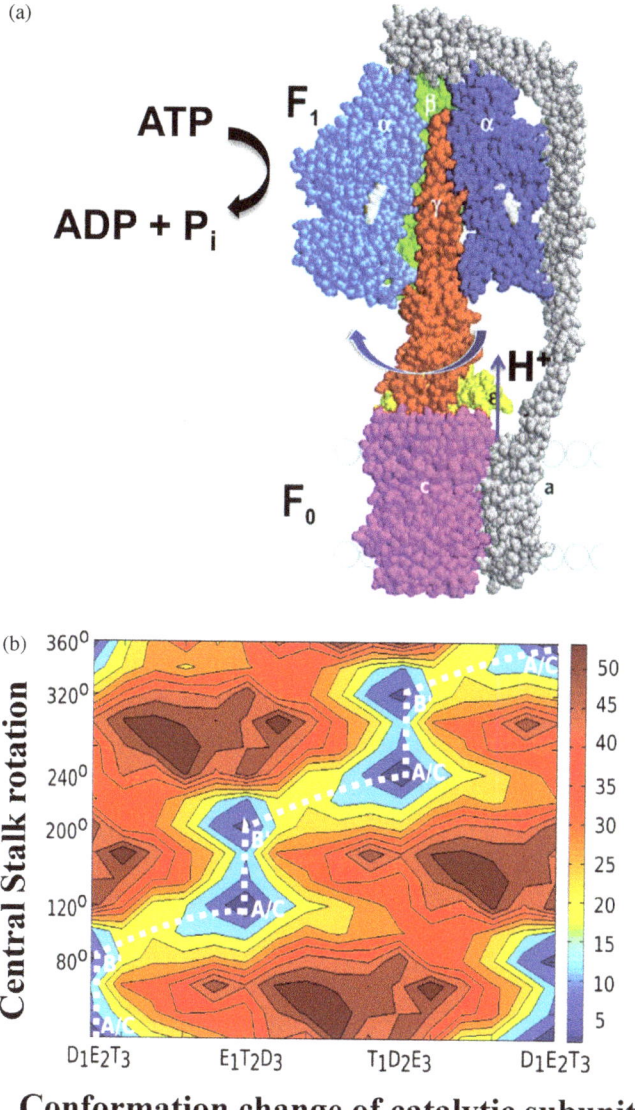

Conformation change of catalytic subunits

Figure 23. (a) F_1/F_0 are two coupled ATPase and ion-pump units that constitute the rotary motor and the stator portions. In the presence of an ion gradient across the membrane the ATP synthesis occurs in F_1. In the opposite direction the ATP hydrolysis occurs while the F_0 acts as an ion pump. (b) The least energy path calculated by our CG model. This path shows the 80°/40° sub-steps observed experimentally. The 80° rotation has a small electrostatic barrier. The chemical step occurs after the 80° rotation. The action of F_0 has also been accounted for by our simulations. Figure b is taken from [84].

respects, this represented closure to my 1978 discussion with Peter Mitchell.

Realizing the power of the CG model, we moved to other systems that operate by the use of chemical energy. In particular, we explored the action of myosins, which are used to generate muscle construction, to carry loads in the cell and for other tasks. A key part of our finding was the generation of free energy landscapes, which helped to establish that the action of myosin is not driven by the so-called "power stroke". A very large part of the community had assumed that myosins are driven in the forward direction by the kinetic energy released in the step where the system moves in what has been described as like the release of a loaded spring. Intuitively, this popular proposal looks very appealing, but it suffers from the same problems as did the enzyme dynamics concept: it ignores the fact that the inertial effect is dissipated quickly before passing the barrier. It also disregards microscopic reversibility, as was pointed out eloquently by our colleague Raymond Astumian [85]. Furthermore, we were able to highlight and resolve the largely ignored but highly important problem of why myosins (e.g., myosin V) move in one direction and not in the opposite direction, despite the fact that the motions in both directions involve similar features [87]. Some scientists' descriptions were analogous to saying that moving in one direction is because the kinetic energy of moving in the given direction keeps pushing the system that way. Of course, this can justify movement in any direction.

Ion Channels

Ion channels are involved in many body actions, including the transfer of information in the brain, and thus probably our emotions, and indeed are targeted by many of the available drugs. I became interested in modeling ion channels in 1983 when I visited the lab of Peter Läuger in Konstanz, a leading physiologist who died in a tragic climbing accident in 1990. I contemplated using my dipolar approach in modeling such channels but did not progress until 1989, when Johan Åqvist, my postdoc at that time, used the Protein Dipoles Langevin Dipole (PDLD) model (that I developed in 1976 and used extensively in different electrostatic studies) to explore the free energy profile for ion transfer trough the gramicidin channel. The same study also included the first all-atom free energy perturbation calculation of the energetics of ion transfer through ion channels. Although this research had an impact on leading experimentalists in the field (e.g., George Eisenman and Bert Sakmann), it had limited impact on theoreticians. Most of those interested in the field were focused on the possible bumps in the free energy profile that could be captured by subsequent potential of mean force calculations, rather than on the much more challenging and important ability to capture the "solvation" of the ion by the channel, in particular, upon moving from water to the mixed environment of induced dipoles, permanent dipoles and water molecules inside the channel.

Despite the intensive collaboration of George Eisenmann with my co-workers, I decided not to spend too much effort clarifying the problems with other studies that were being highly quoted, since there was no available structure of a real ion channel. The situation changed in 1998 when Rod MacKinnon remarkably solved the structure of the potassium channel [88], for which he was awarded the Nobel Prize. On the theoretical front, in a 1999 paper Benoît Roux and MacKinnon [89a] tried to evaluate the energetics of the potassium ion in the channel cavity, using a continuum calculation with a far too low dielectric constant that drastically overestimated the role of the helix dipoles. Then in 2000 Johan Åqvist and Victor Luzhkov published in *Nature* an insightful paper [90] with careful free energy calculations, which elucidated key features of the selectivity filter. Subsequent papers of Roux and colleagues [89b] used visually impressive but still problematic potential of mean force calculations, which worked only in cases of a perfect convergence and could not point at problems in the evaluation of the absolute electrostatic energy, the role of ionizable groups and the effect of the protein polarizability.

At this stage I decided to focus on a deeper understanding of the control of the actual observable, which is the ion current. In 2002 [91] I turned to a Langevin dynamics treatment with a channel model that reflected the PDLD electrostatic profile of the KcsA channel. In a subsequent paper in 2003 [92], we succeeded in reproducing the ion current of K^+ and Na^+, showing that the K^+ ions move with two ions in the selectivity filter, while the Na^+ ions move one by one. More importantly, we actually elucidated the reason for the selectivity, which could be described as having a smaller effective dielectric for charge-charge interaction for the Na^+-Na^+ pair and a larger dielectric for the K^+-K^+ pair. That is, we found that the protein dipoles interact in a stronger way with the smaller Na^+ ions than with the K^+ ions and thus are not able to reorganize so much as in the case of K^+ ions, and this results in a smaller dielectric effect for charge-charge interaction. This important, non-trivial general insight is still not widely appreciated.

In subsequent studies, we strived to understand the microscopic basis of the voltage-current relationships, a challenging task since the available literature focused on macroscopic treatments that did not even consider explicitly the electrodes, nor did they provide a reasonable and physically convincing description of the gating current (which appears at a short time after the change in voltage and before the ions pass through the channel). As in all my previous research, I had to move without clear help from the ion channel literature, so I turned to a microscopic description to confirm my understanding. We built a model that considered explicitly

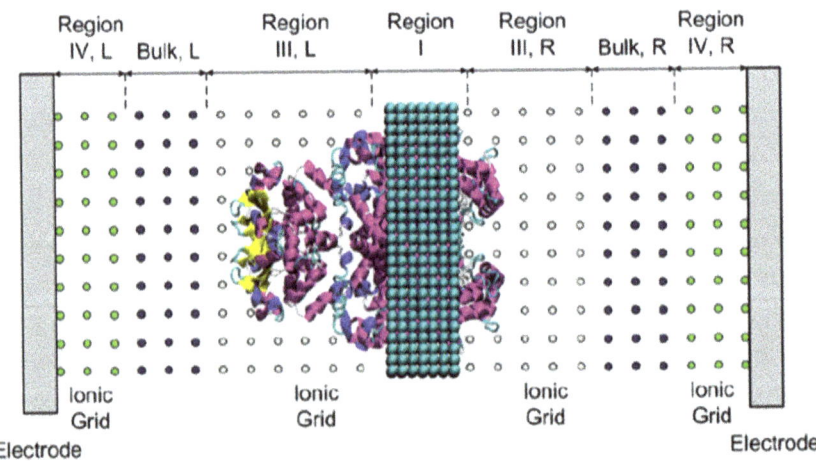

Figure 24.   Our unique model of voltage-activated ion channels. In contrast to other models (including continuum models) we included explicitly the electrolytes and the electrodes. This allowed us to explore with confidence the nature of voltage activation. The figure of the grid model is taken from [93].

the electrolytes in the solution between the membrane and the electrodes. The electrolytes were represented as an explicit grid of charges that respond to the potential from the electrodes, membrane/protein system and the other grid points. After a major verification effort (where we reproduced the known analytically solvable test cases), we were ready to explore many key features of membrane-embedded ion channels. This included quantifying the gating current and the meaning of the measure voltage change in response to charge separation in reaction centers [93, 94]. Our ability to investigate voltage effects explicitly became an important tool in our CG simulations, including a unique modeling of voltage-activated ion channels (Figure 24).

Significantly, we also verified our unique simplified microscopic modeling of ion channels using a Monte Carlo model that considered the electrolytes explicitly with an implicit model of the solvent (this model is called a primitive model) [95]. I have often wondered why the great work of statistical mechanicians in the 80s who used a Monte Carlo primitive model to study ions between electrodes was not extended to electrodes membrane systems.

XIII

The Nobel Prize

O n the night of October 8, 2013, I was very anxious when I went to sleep. I knew that my field was being considered for the Nobel Prize, but I wasn't at all sure about that, or who would be included, if it was. Tami suggested I take half a sleeping pill and, more optimistic than I was, placed flowers at the door for the news crew that might come. At 2 am the phone rang and Tami woke me up and handed it to me. I told her it was too early, since according to my calculations the announcement should have been at 3 am, and I still suspected that the voice on the other end declaring that I had been awarded the 2013 Nobel Prize for Chemistry was a cruel joke. But the voices were Swedish and then I heard the sounds of some of my Swedish friends. I was, of course, thrilled, and we started to prepare for the flood of news media that was expected in a few hours.

The news media arrived starting at 5 am and kept us busy, with our daughter Yael organizing the interviews and photographs (Figure 25). Among other congratulatory calls was one from then-Israeli President Shimon Peres and the Prime Minister at the time, Benjamin Netanyahu. This was followed by a reception and a media event at USC which could claim only one Nobel Laureate 20 years earlier, George Olah, also in chemistry.

My Nobel announcement made a great splash in Israel and the newspapers were full of stories. Reporters went to my kibbutz, Sde Nahum, and interviewed people who knew me, noting that I was the only Nobel Laureate to have been born on a kibbutz. A good friend of my mother related a story about my success. In 1998, she and my mother were talking and my mother said that the schooling in equality leads to mediocrity, whereas research and discoveries are the basis for advanced society. Then

Figure 25.   A photo from the *LA Times* of Tami congratulating Arieh on the morning of the Nobel Prize announcement.

she handed her friend an article of mine in English, telling her it was a major breakthrough in the world of chemistry. And she added, "You will see, one day he will get the Nobel Prize." She wrote as well of a related case, in 1982, when my brother Beni got married and changed his last name to Shalev, wherein my father asked him what he'd do if Arieh gets a Nobel Prize. Beni responded that he'd change his name back to Warshel.

The next two months were full of exciting events, including a trip to Seoul for a high-level *Molecular Frontiers* meeting, where most of the speakers were Nobel Laureates. My invitation in August actually seemed strange to me, until I was told at the meeting that someone promised them a new Nobel Laureate long before it was announced (although I am not sure this story is true).

I was happy to learn that although the prize was generally awarded to "multiscale modeling" it mainly focused on the QM/MM method (as was clear from the Nobel poster) and thus on the 1976 work. Of course, I would have been happier if it was given for computer modeling of enzymes but this was not a realistic political possibility.

The excitement culminated in the Nobel week. Tami and I were received directly on the airplane and then taken to the Grand Hotel in Stockholm, followed by days of exhilarating events for the Laureates and

Figure 26.   The Nobel Prize being awarded to Arieh Warshel by King Carl XVI Gustaf of Sweden. Copyright © Nobel Media AB 2013; Photo: Alexander Mahmoud.

our families, including, of course, the Nobel award ceremony (Figure 26), the amazing banquets at City Hall and the royal palace (Figure 27).

On the third night we were told a "secret", that the next morning we would be woken up by girls and boys dressed in white like angels, in

Figure 27.   Tami and Prince Carl Philips at the Nobel banquet.

celebration of Santa Lucia Day. A previous Nobel Laureate, they shared, was sure he'd died and went to heaven when he was awoken by the little angels. We went along with the fun, and were very pleasantly woken up by the group, who served us special cookies and tea and sang lovely songs. Perhaps this is what happens in heaven, although according to Jewish tradition you are served whale meat and not cookies when you get there.

I lost my voice that week, likely due to a flu I contracted from an Israeli journalist, but fortunately it first struck me only the night after my Nobel lecture. I was taken to Karolinska Hospital, where they gave me the same steroid treatment they give opera singers with the same condition. Over the following week I gave a series of lectures at various Swedish universities with the following routine: I would have a reasonable voice to lecture, but I'd lose my voice again at the evening parties. Interestingly, at every university the local experts gave me a different anti-flu medication.

One of the Nobel events was the BBC's "Nobel Minds" program, where we were interviewed about our next steps. What would we do with our fame, for instance. Most of the Laureates talked of their plans to change the world with various social activities. Economics Laureate Eugene Fama and I answered that we would rather continue to do what we know how to do in our respective fields.

Joining us in Stockholm for the Nobel week were many family members, including Merav, her husband Rafi and their kids, Maya, Daniel and Neev; Yael; Tami's sister Tova, her husband Danny and two of their daughters Amit and Nitzan, and Tami's cousin Hanna; my brother Beni and his wife Orit; and our long-time friend Ruth Sharon. Everyone had a great time and went home with unforgettable memories.

XIV

Life After the Nobel Prize

The Nobel Prize dramatically enhanced my status. Not insignificantly, part of the reward has been that people stopped questioning my ideas. However, this was a problem in my lectures, since without challenging questions I could not clarify my ideas. Similarly, I also got much less criticism.

The Nobel excitement has never fully subsided. I am routinely invited to major events around the world, including forums, institution openings and keynote lectures. Tami and I have been treated as celebrities, courted with first-class flights, limousines, hotel suits and more. And it has been thrilling to give lectures in countries where they adore Nobel Laureates, seeing enormous lines of kids waiting for a signature or a selfie.

Around 2015 I was invited to the Chinese University of Hong Kong in Shenzhen (CUHKSZ) to launch an institute for biological modeling, which was inaugurated in 2017, with the building completed in 2018. Named the Warshel Institute of Computational Biology, it was set up with sizable space and a relatively large research group that has been focusing on molecular modeling and bioinformatics.

G-Proteins Coupled Receptors

Despite the new load of public activity, I continued very active research, with a dynamic group and a constant flow of publications. Our research included a continuation of the CG simulations, extending them to modeling the action of G-protein coupled receptors (GPCR). The amazing GPCR systems, which control the cell response to different stimuli, involve a membrane protein that spans the membrane, with an external side that can bind various ligands (e.g., key hormones) and an inner membrane side that binds a G-protein with a bound GDP. When the proper ligand binds to the external part the receptor is activated and the G-protein exchanges the GDP for a GTP. This activates the specific intra-cellular system to perform its specific task.

Major studies that culminated in the structural elucidation of GPCRs provided a molecular view of the activation of these systems and challenged simulation studies to generate the corresponding structure-energy relationships. All-atom MD simulations that used powerful computers have provided instructive trajectories of the activation process, but still could not yield reliable free energy landscapes. Our CG approach allowed us to generate approximate landscapes that provide an energy-based description of the activation process (see Figure 28). Our strategy has

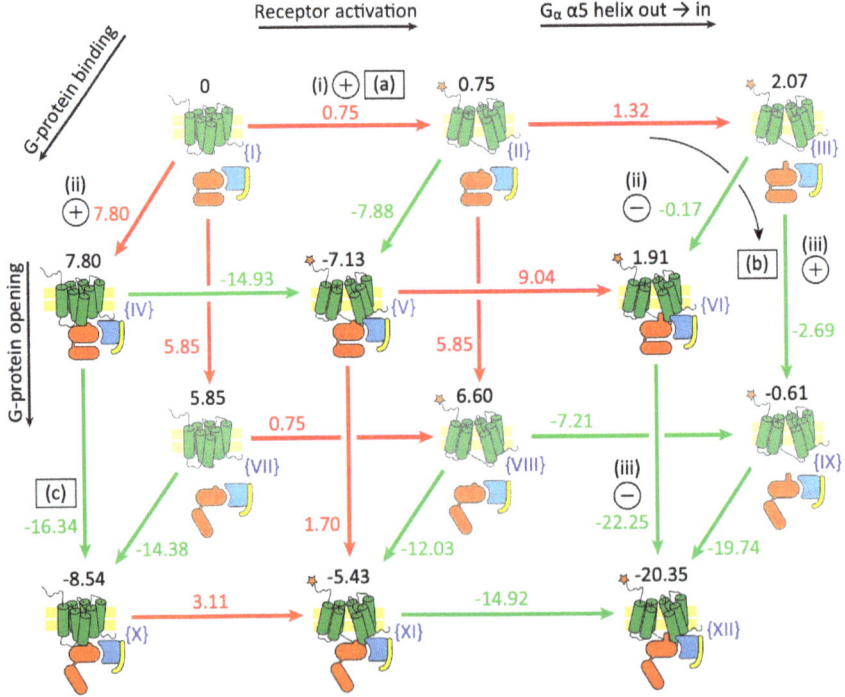

Figure 28.   The energetics of the different endpoints on the landscape of the activation of the β-2 adrenergic receptor (β2AR). Upon binding of adrenaline, the GPCR undergoes conformational changes that result eventually in the release of GDP from the G-protein and binding of GTP. The energy values were determined by our CG model. The figure is taken from [96].

been applied to the analysis of the action of the adrenaline receptor, the glucagon-like peptide 1 (GLP-1) hormone receptor [97] and the μ-opioid receptor [98].

Computer-Aided Enzyme Design

Although our ability to predict and reproduce the catalytic power of enzymes has been very persuasive for some researchers, it has left a large part of the community unconvinced, in part because of the many other proposals described in textbooks. It seems that the only way to convince some people is to perform miracles, such as designing powerful enzymes from scratch. The most visible attempts to move in this direction involved so called "computer-aided enzyme design", which was not based on the ability to reproduce enzyme catalysis, but on the ability to generate enzyme structures even if they have only small catalytic effects. This type of approach received great exposure in high-impact journals. However, a large part of the alleged catalytic effect was due to choosing as a reference the reaction in dilute solutions at neutral pH solution, rather than in 55M of the reacting groups. Considering our perspective on enzyme modeling, we felt compelled to point out that the minor success of this enzyme design approach was almost entirely owed to the collaboration with experimentalists, who succeeded in refining the rate acceleration by directed evolution, although the acceleration was much smaller than the catalysis obtained by naturally evolved enzymes.

This gap in full acceptance of our approach put significant pressure on us to demonstrate that we could do better. My interest in enzyme design had started much earlier, in 1984, when I suggested to Alan Fersht, who was among the most important pioneers of enzyme engineering, that we study the results of mutations of t-RNA synthetase. His response was that I send him our predictions and he'll see if I'm right. My fear, I wrote back, was he would announce it when I was wrong. Eventually, in 1987, we made some successful predictions about mutations in subtilisin [99], and in subsequent years reproduced the effects of many observed mutations [100]. Still, we could not claim that we designed a better enzyme. Around 2010 we felt we should enter the enzyme design "contest", and we focused on the catalysis of Kemp eliminase, the center of a series of

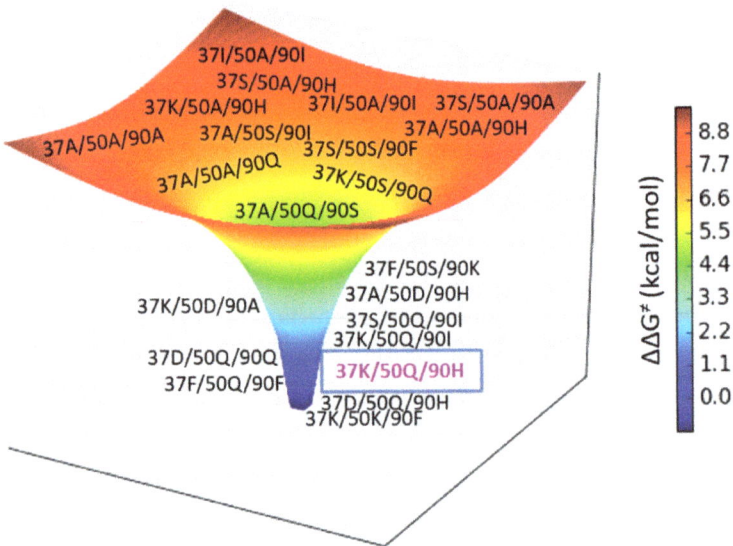

Figure 29.   Screening mutations of Kemp eliminase by simulating directed evolution. The figure depicts a ranking of the number of possible three-mutation combinations according to the prediction of the corresponding transition state stabilization. The color scaling indicates the extent of stabilization (red, least stable; dark blue, most stable) achieved by introducing mutations at three positions (Q37, H50, D90), using different types of amino acids. The flat top part of the 3D diagram corresponds to a large number of mutation combinations with a high $\Delta\Delta G^{\ddagger}$ value, whereas the narrow bottom indicates a very low number of mutations leading to a low $\Delta\Delta G^{\ddagger}$ value (and thus strong catalytic power). The experimentally found mutation combination (Q37K/H50Q/ D90H) is highlighted (blue box). The figure is taken from [103].

papers in prestigious journals. We gradually found that it is hard to generate large catalytic activity with this enzyme (due to the delocalized charge distribution of the transition state), but we could reproduce the observed results of directed evolution, provided we had the structure of the different mutants [101, 102]. We also tried to follow the path of directed evolution and found out that we can evaluate the catalysis in a known active site structure. However, we had difficulties in generating the effect of residues that may not be in the active site on the preorganization of the active site residues. Our effort of progressing on this front involved automatic generation of rotamers that succeeded in creating a catalytic funnel, where mutants with actual significant catalysis are found at the bottom of the funnel (Figure 29) [103].

In addtion to the projects mentioned above, we made significant advances in various medically related directions, including: the molecular origin of cancer, drug resistance, HIV and the Covid-19 epidemic.

I strongly believe that rational computer-aided enzyme design will become quantitative at some stage in the future, but we can't predict how soon that will be.

XV

Epilogue

Some Personal Observations

As I look back on my career, some observations based on my cumulative experience come to mind. My advances were based on early use of computers and on intuition rather than step-by-step systematic derivations. Using computers at a time when numerical solutions looked inferior to elegant analytical solutions allowed me to visit complex problems long before most researchers. This capacity to explore many complex systems gave me confidence in the general validity of my approach. This included, in the case of enzymes, investigating a host of alternative proposals, rather than leaving some options as open possibilities. Of course, jumping in different directions and declaring confidence in the corresponding results made some researchers suspicious, wondering how other established scientists hadn't reached the same conclusions.

An instructive but far from unique example was David Chandler and co-workers' attempt in 1993 to model the primary event in photosynthesis and publishing a *JACS* paper [104] declaring that the conclusions reached by Parson and me were wrong. However, as was the case with Schulten and Karplus, he hadn't included water in and around the protein. Candler had also neglected the "self-energies" the components would have if they were separated to infinite distance in a dielectric medium.

Considering the topic's importance, we wrote a detailed response paper [105], pointing out the extremely serious errors in Chandler's work.

As expected, he got very upset (despite having been the initial offender) and attacked our Langevin dipoles model, which he called "imaginary water". This led us to demonstrate that the Langevin dipole model represents the physics much more accurately than not including water at all (a treatment that can lead to errors of more than 50 kcal/mol). Chandler then wondered why those he considered leaders in the field were not using related models and do not subscribe to our dielectric concepts. We were compelled to clarify that those authorities had little experience in validating and evaluating the energies of charges in proteins, which happened to be my main focus since 1976 (after identifying the crucial role of electrostatics in proteins). Of course, this did not go well with Chandler, who was a world leader in studying the properties of water, and we were not very successful in teaching him about the energies of charges in proteins and solutions. The point is that identifying the crucial problem early and trying to examine how to treat it is a critical guide, and more important than gaining rigorous training in unrelated subjects.

Importantly, even as I relied on computer simulations to verify intuitive hints and present an "empirical" test of the models I used, I had to establish the rigor of my approaches and respond to the anticipated criticism that my findings were not really valid. A more conventional scientific path builds a case by publishing numerous incremental papers that examine different effects and thus looks like the basis of solid validation. Similarly, many popular approximations have been accepted because they started from a rigorous treatment and only introduced the key assumptions at a later stage, after losing the attention of the critical reader. However, in dealing with complex systems, many of those "validations" involve effects that have very little to do with the relevant issues (e.g., exploring thermal atomic motion is not particularly relevant to the determination of the activation barrier in an enzyme). Instead, I focused directly on the key effect and in parallel examined my assumptions, but did not generate many papers on the corresponding validation process, which I believed was too trivial to deserve a detailed article. In fact, Mike Levitt told me once that I work like a commando, while others operate like an army. I have, of course, written long papers proving the rigor of my approach, but they were rarely read by those who presumed my research was based on off-the-wall ideas. Rudy Marcus told me once that apparently there are two Warshels, one who writes long, mostly unread rigorous papers and another who starts with untested hypotheses.

Interestingly, all this made it tempting to dismiss my proposals by presenting me as a "wild swashbuckler" scientist. A 1991 article by Anne Moffat about the field of molecular modeling in the NSF publication *Mosaic*, for example, claimed I believed that all the catalytic power of enzymes is associated with electrostatic effects, while Karplus and Andrew McCammon, who took a more "Catholic approach" that includes molecular dynamics and free energy calculations, concluded that other effects could also be important. The piece then went on about QM/MM calculations of enzymes. The problem was that at the time of the interview I was the only one who consistently used molecular dynamics and free energy calculations as well as QM/MM modeling (in particular the EVB) in realistic studies of enzyme catalysis. These studies led to my conclusions about the importance of electrostatic effects, which only came after checking all the other relevant options. It seems to me that Moffat wasn't negatively inclined, but simply could not believe that my assertions of professionalism could be real given the simplicity of my final conclusions about electrostatic effects.

Another, more convoluted example was a story in *Nature* by Philip Ball [106] that explored the issue of enzyme tunneling, wondering whether it was "by chance or by design". In the first draft that Ball sent me, he wrote that "Warshel believes that the idea that tunneling helps catalysis is a red herring." However, he did not mention that at that time we were the only group exploring this idea consistently with actual simulations. He also cited calculations of very large tunneling in the enzyme lipoxygenase (where the calculations did not include the actual enzyme), while not mentioning that my colleagues and I had already published quantitative calculations of the tunneling contributions in this and other enzymatic reactions. I wrote to Ball that it looked like he wanted to present me like some kind of mad scientist, but I was actually the only one in his article who had investigated the tunneling idea using real simulations. I also sent him a photo of my path integral simulation of quantum tunneling in carbonic anhydrase. The final version of the article included the photo, but without sourcing it as coming from my simulations. Nor did it make note that I actually based my arguments on careful simulations, rather than presumably just on an opinion I might have had. Of course, this may have helped make the story more intriguing, but it didn't provide the reader with the actual scientific facts.

In some respects I felt like those early explorers whose claims of discoveries were questioned. My early reputation at USC illustrates this nicely. I tried to point out politely in various seminars that I had studied the subject that the speaker was discussing and had already moved much further ahead, which led, of course, to whispers by my colleagues. Then, a former postdoc of Anna Krylov, also a Chemistry professor at USC, told my postdoc in an ACS meeting that she had always thought my comments were highly exaggerated, "but now I found out that they are all actually true."

At this point I'd like to say a word about scientific credit. That is, when one is involved in innovative and competitive science, concern may arise about receiving credit for various discoveries. Credit comes to some scientists naturally, being at prestigious institutions; in many cases they even get it for work pioneered by lesser-known scientists. Others have to try hard to get recognition for their actual contributions.

For some time, I thought my papers were not credited because of my less-than-optimal English. I recalled the great physicist Max Delbrück (who in 1978 accepted the request of Martin Kamen to edit my *PNAS* paper on rhodopsin [107]) complaining, when I sat near him, that "this bum can't even write English." However, gradually I learned that it had little to do with my writing. Mike and my paper on folding is a good example. Mike wrote the article very eloquently, but it went significantly underquoted in cases where it was supposed to be a key reference. Similarly I even learned about some researchers who were threatened that if they cited our 1976 QM/MM paper, they might not get tenure.

I was inclined to relax my concerns when I heard Max Perutz saying that at the end, truth will always emerge. But I feared that this might not happen when it most counts, while I was still alive. I must mention that Max told me once not to worry so much that someone would steal my ideas, since they wouldn't understand them anyway. While I knew he was right, I didn't stop worrying. I even published one of my best ideas, about moving from a simplified folding model to an all-atom model, in a journal that almost no one reads so my approach wouldn't be stolen. For that matter, no one noticed this idea for a long time after its publication in a more widely available journal either.

I often wondered what could help disseminate my ideas most effectively, but I couldn't find the optimal path, in particular considering the fact that some leading researchers had done their best not to mention my work. Once, I almost found the solution. After seeing that my electrostatic ideas were systematically ignored or adopted without credit or attributed to others who missed the main points, I tried to repeat the same arguments in different ways in different publications. This strategy became particularly interesting when I found that the discussion of the challenge in predicting pK$_a$s was constantly attributed to a group of researchers who missed the main point — that it is to get the correct prediction for cases with very large deviations from the value in solution (since groups on the surface of proteins have very similar pK$_a$s to the corresponding pK$_a$ in water). As the scientist who started microscopic pK$_a$ calculations in proteins and argued the relevant points in numerous papers, I was not sure what to do. In 2001 I wrote a short piece with my graduate student Nili Schutz titled "What are the Dielectric 'Constants' of Proteins and How to Validate Electrostatic Models?" [108] where slowly, step by step, we reviewed the main issues, pointing out again that the dielectric for charge-charge interactions in proteins is much larger than what is assumed by most researchers. To my surprise, this article began to be widely quoted. I am still not sure what the effect of being cited is, considering Martin Kamen's remark that "they only quote you when you are wrong." Nevertheless, eventually my H index for quotations became quite high.

My issues with publishing were not unique to me, and appeared to be common to other researchers who did not publish incremental papers and challenged current dogma. In my case, my nearly constant problems with my articles could be traced to the reviewing system, where reviewers can mislead their editors without facing a penalty for presenting incorrect information. My proposal to correct this — if the author can demonstrate that one or two of the reviewer's claims is intentionally misleading by proving prior knowledge of problems with those claims, the reviewer should be disqualified — doesn't have a great chance of being adopted. This issue has been a particular problem for me, since many of my submissions to high-impact journals were rejected by reviewers maintaining that no one had ever proposed the point that I was attacking (e.g., saying that dynamics is an important factor in enzyme catalysis). In one case, when I told the editor of *Science* that the reviewer was simply lying, I was told not to use the "L" word. It didn't help that I provided countless quotations from the corresponding paper. This led me to start some of my papers with the relevant citations. Some of my friends used to laugh at me and say that I must enjoy fighting with reviewers, but I really felt it was mostly a completely useless effort, since the editors frequently accepted the reviewers' point of view.

Once, at a relatively advanced stage of my career, Rudy Marcus told me, "Arieh, you haven't yet found a journal that will publish your papers without problems." This was useful advice, but only at the stage where I became sufficiently known and my papers were noticed regardless of where they were published.

My attempts to get credit for my work led to my interest in having methods named after me. Yet despite inventing numerous methods and assigning them their own names, I had no success getting the name Warshel attached to them. Often the names I had chosen were used to downgrade the methods, for example, by emphasizing that the Empirical Valence Bond is "empirical", or that the Frozen Density Functional is "frozen". According to my daughter Yael, the communication professor, the problem might have been in my choice of names. She had observed that in TV advertising, the first name chosen for a product is usually not the best name.

Some readers might believe that science is pure, and the search for truth overrides any other consideration. However, as in other human endeavors, competition, ego and ambition play a major role. As illustration, in 2013 USC invited the great Israeli author Meir Shalev to talk about his work. During the discussion he mentioned that he is most proud of a prize he received from the Entomological Society of Israel for his exact description of the world of insects. In contrast to regular literature prizes, he explained, this prize was not based on any intrigue or political consideration since it was awarded by scientists. It was hard for me to refrain from correcting his impression about scientific purity, but I only smiled.

Receiving the Nobel Prize, considered as the ultimate scientific recognition, makes one wonder what led to the honor. Of course, the actual discovery underlying the award must be very important, but being in the right place at the right time is also a major factor. In my case, being at Shneior Lifson's group when computers just began to be used in molecular modeling, and my time at the MRC, where so many pioneering discoveries were made, must have helped.

It is also interesting to reflect on how I conducted my scientific research, which was influenced by my experience at the Weizmann Institute and the MRC. In Shneior's group, I always worked alone or with a few collaborators. At the MRC, I witnessed great discoveries being made by small research groups and by scientists who dedicated many years to a single major project. I continued with this tradition at USC, trying to push my projects by myself, and sometimes with one or two co-workers. I realized, however, that I could not ask for grant support without having a group to support. In fact, in the early 80s, I attended a CECAM workshop in Orsay, outside of Paris, where the workshop director, Carl Moser, advised me to get a real group if I was going to make it in the United States. I told him I feared that would only slow me down, but I also came to understand that funding was contingent on working with a group. Eventually having a group, although a relatively small one, has been greatly beneficial, and I have worked with some very talented researchers.

The Future of Biological Simulations

Looking at the progress of multiscale molecular simulations, one may wonder what the future holds. As a start, it is clear we are not yet at the stage of having sufficiently accurate results to provide quantitative predictions. A case in point is the situation with high-level *ab initio* quantum mechanical calculations of very small (e.g., triatomic) molecules. Such calculations moved over the last 40–50 years from being very qualitative to being so quantitative that experimental spectroscopists are using them with full confidence when analyzing experimental results.

I recall that in a meeting in Amsterdam in 1990 Peter Kollman and Bill Godard had a bet about when the folding problem would be solved — in 10 years or 30 years. Yet even now we do not have a fully predictive solution for this challenge, although microscopic simulations of the fast folding of small proteins are getting better, and machine-learning approaches seem to be giving promising results. Similarly, a few years ago, at the end of a meeting about creating a research group that would focus on cell modeling, Ray Stevens asked people to write on the blackboard how long it would take to have full atomistic modeling of an entire cell. Several participants wrote three years; I wrote it would take over 70 years for correct microscopic calculations. The clock is still ticking, but a reasonable CG modeling should provide a useful picture much earlier.

We might also ask when we will be able to make sufficiently quantitative predictions of functional properties of proteins. Here we note that for questions such as the origin of the catalytic power of enzymes, our approaches are often reasonable in many cases, since the observed effects are very large. However, for enzyme design and for binding calculations it is important to go below the 1 kcal/mol error limit, and this still presents an enormous challenge. The source of the current errors may be due to convergence difficulties or to the quality of the parameters of the potential surfaces. As we found in trying to follow directed evolution with our simulation approaches, it is also crucial to generate many protein configurations. Obviously, it is important to overcome this sampling problem.

For now, it is unclear how long it will take to reach fully predictive simulations of functional properties. It is also uncertain whether the crucial advances will come from much stronger computer power or from algorithmic advances. However, in view of the constant progress in the field, we will definitely get to the point of routine simulations that provide quantitative results and guide studies of biological systems. In fact, it is evident to me that we will have dedicated computerized machines that will guide enzyme and drug design.

Another direction that might provide effective advances will be artificial intelligence and big data. This is in some respects orthogonal to our deductive approach, as it is based on the knowledge of related systems rather than on physical principles. However, predictions based on similarity have significant potential. One might further argue that using many known protein structures to deduce the effective interaction between different residues is similar to a statistical derivation of potential functions.

Regardless of the speed of advances in computer-aided enzyme design, I feel that I have been blessed by being so deeply involved in the initiation of the emerging field of computer modeling of biological molecules. No less was my fortune in having such a fascinating path from the kibbutz to the Nobel Prize. That path was forever enriched by the joy of discoveries driven by endless curiosity and interest in creating previously unimaginable new ways of viewing nature. Playing curiously as a child with a cat apparently spurred me on to continue investigating what underlies it and other lifeforms.

References

1. Bettelheim, B. (1969). *The Children of the Dream*, Macmillan, New York.
2. Gilad, Z. & Meged, M. (1953). *Sefer ha-Palmaḥ* (The Palmach Book), Hakibbuz Hameuchad edition, Tel Aviv.
3. Alterman, N. (1957). *City of the Dove*, Hakibbuz Hameuchad edition, Tel Aviv.
4. Young, H., Freedman, R., Sears, F. & Zemansky, M. (1949). *University Physics*, Pearson Education, London.
5. Merzbacher, E. (1970). *Quantum Mechanics* (2nd edition), John Wiley & Sons, New York.
6. Lifson, S. & Warshel, A. (1968). A Consistent Force Field for Calculation on Conformations, Vibrational Spectra and Enthalpies of Cycloalkanes and n-Alkane Molecules. *Journal of Chemical Physics* 49, 5116–5129.
7. Born, M. & Huang, K. (1954). *Dynamical Theory of Crystal Lattices*, Clarendon Press, Oxford.
8. Kuan, T. S., Warshel, A. & Schnepp, O. (1970). Intermolecular potentials for N_2 molecules and lattice vibrations of solid alpha-N_2. *Journal of Chemical Physics* 52, 3012.
9. Schnepp, O. & Jacobi, N. (1972). The lattice vibrations of molecular solids. In: Prigogine, I. & Rice, S. A. (eds.), *Advances in Chemical Physics*, Vol. 22, John Wiley & Sons, New York.
10. Warshel, A. & Bromberg, A. (1970). Oxidation of 4a,4b-dihydrophenanthrenes. III. A theoretical study of the large kinetic isotope effect of deuterium in the initiation step of the thermal reaction with oxygen. *Journal of Chemical Physics* 52, 1262.
11. Levitt, M. & Lifson, S. (1969). Refinement of protein conformations using a macromolecular energy minimization procedure. *Journal of Molecular Biology* 46, 269–279.

12. Warshel, A., Levitt, M. & Lifson, S. (1970). Consistent force field for calculation of vibrational spectra and conformations of some amides and lactam rings. *Journal of Molecular Spectroscopy* 33, 84–99.
13. Tully, J. C. & Preston, R. K. (1971). Trajectory surface hopping approach to nonadiabatic molecular collisions — reaction of H$^+$ with D$_2$. *Journal of Chemical Physics* 55, 562.
14. Warshel, A. & Karplus, M. (1972). Calculation of ground and excited-state potential surfaces of conjugated molecules. 1. Formulation and parametrization. *Journal of the American Chemical Society* 94, 5612–5625.
15. Warshel, A. & Huler, E. (1974). Theoretical evaluation of potential surfaces, equilibrium geometries and vibronic transition intensities of excimers — pyrene crystal excimer. *Chemical Physics* 6, 463–468.
16. Warshel, A. & Shakked, Z. (1975). Theoretical study of excimers in crystals of flexible conjugated molecules — excimer formation and photodimerization in crystalline 1,4-diphenylbutadiene. *Journal of the American Chemical Society* 97, 5679–5684.
17. Levitt, M. & Warshel, A. (1975). Computer simulation of protein folding. *Nature* 253, 694–698.
18. Warshel, A. (1977). Energy-structure correlation in metalloporphyrins and the control of oxygen binding by hemoglobin. *Proceedings of the National Academy of Sciences of the United States of America* 74, 1789–1793.
19. Böttcher, C. J. F. (1971). *Theory of Electric Polarization*, Elsevier, Amsterdam.
20. Warshel, A. & Levitt, M. (1976). Theoretical studies of enzymic reactions — dielectric, electrostatic and steric stabilization of carbonium-ion in reaction of lysozyme. *Journal of Molecular Biology* 103, 227–249.
21. Warshel, A. & Karplus, M. (1975). Semiclassical trajectory approach to photoisomerization. *Chemical Physics Letters* 32, 11–17.
22. Warshel, A. (1976). Bicycle-pedal model for 1st step in vision process. *Nature* 260, 679–683.
23. Frutos, L. M., Andruniow, T., Santoro, F., Ferre, N. & Olivucci, M. (2007). Tracking the excited-state time evolution of the visual pigment with multiconfigurational quantum chemistry. *Proceedings of the National Academy of Sciences of the United States of America* 104, 7764–7769.
24. Warshel, A. (1977). Interpretation of resonance raman-spectra of biological molecules. *Annual Review of Biophysics and Bioengineering* 6, 273–300.

25. Warshel, A. (1979). Calculations of chemical processes in solutions. *Journal of Physical Chemistry* 83, 1640–1652.

26. Perutz, M. F. (1978). Electrostatic effects in proteins. *Science* 201, 1187–1191.

27. Warshel, A. & Weiss, R. M. (1980). An empirical valence bond approach for comparing reactions in solutions and in enzymes. *Journal of the American Chemical Society* 102, 6218–6226.

28. Storm, D. R. & Koshland, D. E. (1972). Indication of the magnitude of orientation factors in esterification. *Journal of the American Chemical Society* 94, 5805–5814.

29. Arieh, W. (1982). Dynamics of reactions in polar solvents. Semiclassical trajectory studies of electron-transfer and proton-transfer reactions. *Journal of Physical Chemistry* 86, 2218–2224.

30. Heller, E. J. (1981). Frozen Gaussians: A very simple semiclassical approximation. *Journal of Chemical Physics* 75, 2923.

31. Hwang, J. K. & Warshel, A. (1985). Semiclassical simulations of the spectra of anharmonic molecules — problems and alternatives. *Chemical Physics Letters* 115, 281–285.

32. Warshel, A. & Hwang, J. K. (1985). Quantized semiclassical trajectory approach for evaluation of vibronic transitions in anharmonic molecules. *Journal of Chemical Physics* 82, 1756–1771.

33. Warshel, A. & Hwang, J. K. (1986). Simulation of the dynamics of electron-transfer reactions in polar solvents — semiclassical trajectories and dispersed polaron approaches. *Journal of Chemical Physics* 84, 4938–4957.

34. Warshel, A. & Russell, S. T. (1984). Calculations of electrostatic interactions in biological-systems and in solutions. *Quarterly Reviews of Biophysics* 17, 283–422.

35. Warshel, A. (1984). Simulating the energetics and dynamics of enzymatic reactions. In: Pullman, A. & Chagas, C. (eds.), *Specificity in Biological Interactions*, Springer, Dordrecht.

36. Kubo, R. & Toyozawa, Y. (1955). Application of the method of generating function to radiative and non-radiative transitions of a trapped electron in a crystal. *Progress of Theoretical Physics* 13, 160–182.

37. Warshel, A., Stern, P. S. & Mukamel, S. (1983). Semiclassical calculation of electronic spectra of supercooled anharmonic molecules. *Journal of Chemical Physics* 78, 7498–7500.

38. Warshel, A. & Parson, W. W. (2001). Dynamics of biochemical and biophysical reactions: insight from computer simulations. *Quarterly Reviews of Biophysics* 34, 563–679.

39. Warshel, A. (1980). Role of the chlorophyll dimer in bacterial photosynthesis. *Proceedings of the National Academy of Sciences of the United States of America* 77, 3105–3109.

40. Warshel, A. & Parson, W. W. (1987). Spectroscopic properties of photosynthetic reaction centers. 1. Theory. *Journal of the American Chemical Society* 109, 6143–6152.

41. Warshel, A., Creighton, S. & Parson, W. W. (1988). Electron-transfer pathways in the primary event of bacterial photosynthesis. *Journal of Physical Chemistry* 92, 2696–2701.

42. Creighton, S., Hwang, J. K., Warshel, A., Parson, W. W. & Norris, J. (1988). Simulating the dynamics of the primary charge separation process in bacterial photosynthesis. *Biochemistry* 27, 774–781.

43. Treutlein, H., Niedermeier, C., Schulten, K., Deisenhofer, J., Michel, H., Brünger, A. & Karplus, M. (1988). Molecular dynamics simulation of the primary processes in the photosynthetic reaction center of rhodopseudomonas viridis. In: Pullman, A., Jortner, J. & Pullman, B. (eds.), *Transport through Membranes: Carriers, Channels and Pumps*, Springer, Dordrecht.

44. Warshel, A., Chu, Z. T. & Parson, W. W. (1989). Dispersed polaron simulations of electron-transfer in photosynthetic reaction centers. *Science* 246, 112–116.

45. Hwang, J. K. & Warshel, A. (1997). On the relationship between the dispersed polaron and spin-boson models. *Chemical Physics Letters* 271, 223–225.

46. Warshel, A. (1991). *Computer Simulation of Chemical Reactions in Enzymes and Solutions*, John Wiley & Sons, New York.

47. Kamerlin, S. C. L., Sharma, P. K., Prasad, R. B. & Warshel, A. (2013). Why nature really chose phosphate. *Quarterly Reviews of Biophysics* 46, 1–132.

48. Langen, R., Schweins, T. & Warshel, A. (1992). On the mechanism of guanosine triphosphate hydrolysis in ras p21 proteins. *Biochemistry* 31, 8691–8696.

49. Shurki, A. & Warshel, A. (2004). Why does the Ras switch "break" by oncogenic mutations? *Proteins: Structure, Function and Bioinformatics* 55, 1–10.

50. Prasad, B. R., Plotnikov, N. V., Lameira, J. & Warshel, A. (2013). Quantitative exploration of the molecular origin of the activation of GTPase. *Proceedings of the National Academy of Sciences of the United States of America* 110, 20509–20514.

51. Warshel, A. (1979). Conversion of light energy to electrostatic energy in the proton pump of halobacterium-halobium. *Photochemistry and Photobiology* 30, 285–290.

52. Warshel, A., Chu, Z. T. & Hwang, J. K. (1991). The dynamics of the primary event in rhodopsins revisited. *Chemical Physics* 158, 303–314.

53. Warshel, A. & Chu, Z. T. (2001). Nature of the surface crossing process in bacteriorhodopsin: Computer simulations of the quantum dynamics of the primary photochemical event. *Journal of Physical Chemistry B* 105, 9857–9871.

54. Pauling, L. (1946). Molecular architecture and biological reactions. *Chemical and Engineering News* 24, 1375–1377.

55. Hwang, J. K., King, G., Creighton, S. & Warshel, A. (1988). Simulation of free-energy relationships and dynamics of Sn2 reactions in aqueous solution. *Journal of the American Chemical Society* 110, 5297–5311.

56. Warshel, A., Sharma, P. K., Kato, M., Xiang, Y., Liu, H. B. & Olsson, M. H. M. (2006). Electrostatic basis for enzyme catalysis. *Chemical Reviews* 106, 3210–3235.

57. Warshel, A. & Florian, J. (1998). Computer simulations of enzyme catalysis: Finding out what has been optimized by evolution. *Proceedings of the National Academy of Sciences of the United States of America* 95, 5950–5955.

58. Warshel, A., Strajbl, M., Villa, J. & Florian, J. (2000). Remarkable rate enhancement of orotidine 5'-monophosphate decarboxylase is due to transition-state stabilization rather than to ground-state destabilization. *Biochemistry* 39, 14728–14738.

59. Jencks, W. P. (1987). *Catalysis in Chemistry and Enzymology*, Dover, New York.

60. Warshel, A., Florian, J., Strajbl, M. & Villa, J. (2001). Circe effect versus enzyme preorganization: what can be learned from the structure of the most proficient enzyme? *ChemBioChem* 2, 109–111.

61. Richard, J. P. & Amyes, T. L. (2004). On the importance of being zwitterionic: enzymatic catalysis of decarboxylation and deprotonation of cationic carbon. *Bioorganic Chemistry* 32, 354–366.

62. Gillan, M. J. (1987). Quantum classical crossover of the transition rate in the damped double well. *Journal of Physics C: Solid State Physics* 20, 3621–3641.

63. Hwang, J. K., Chu, Z. T., Yadav, A. & Warshel, A. (1991). Simulations of quantum-mechanical corrections for rate constants of hydride-transfer reactions in enzymes and solutions. *Journal of Physical Chemistry* 95, 8445–8448.

64. Hwang, J. K. & Warshel, A. (1993). A quantized classical path approach for calculations of quantum-mechanical rate constants. *Journal of Physical Chemistry* 97, 10053–10058.

65. Hwang, J. K. & Warshel, A. (1996). How important are quantum mechanical nuclear motions in enzyme catalysis? *Journal of the American Chemical Society* 118, 11745–11751.

66. Kamerlin, S. C. L., Mavri, J. & Warshel, A. (2010). Examining the case for the effect of barrier compression on tunneling, vibrationally enhanced catalysis, catalytic entropy and related issues. *FEBS Letters* 584, 2759–2766.

67. Kamerlin, S. C. L. & Warshel, A. (2010). An analysis of all the relevant facts and arguments indicates that enzyme catalysis does not involve large contributions from nuclear tunneling. *Journal of Physical Organic Chemistry* 23, 677–684.

68. Karplus, M. & McCammon, J. A. (1981). The internal dynamics of globular-proteins. *CRC Critical Reviews in Biochemistry* 9, 293–349.

69. Careri, G., Fasella, P. & Gratton, E. (1979). Enzyme dynamics: the statistical physics approach. *Annual Review of Biophysics and Bioengineering* 8, 69–97.

70. Wolf-Watz, M., Thai, V., Henzler-Wildman, K., Hadjipavlou, G., Eisenmesser, E. Z. & Kern, D. (2004). Linkage between dynamics and catalysis in a thermophilic-mesophilic enzyme pair. *Nature Structural & Molecular Biology* 11, 945–949.

71. Klinman, J. P. (2006). Linking protein structure and dynamics to catalysis: the role of hydrogen tunnelling. *Philosophical Transactions of the Royal Society B-Biological Sciences* 361, 1323–1331.

72. Henzler-Wildman, K. A., Lei, M., Thai, V., Kerns, S. J., Karplus, M. & Kern, D. (2007). A hierarchy of timescales in protein dynamics is linked to enzyme catalysis. *Nature* 450, 913–916.

73. Warshel, A. (1984). Dynamics of enzymatic reactions. *Proceedings of the National Academy of Sciences of the United States of America* 81, 444–448.

74. Pisliakov, A. V., Cao, J., Kamerlin, S. C. L. & Warshel, A. (2009). Enzyme millisecond conformational dynamics do not catalyze the chemical step. *Proceedings of the National Academy of Sciences of the United States of America* 106, 17359–17364.

75. Kamerlin, S. C. L. & Warshel, A. (2010). At the dawn of the 21st century: Is dynamics the missing link for understanding enzyme catalysis? *Proteins: Structure, Function and Bioinformatics* 78, 1339–1375.

76. Page, M. I. & Jencks, W. P. (1971). Entropic contributions to rate accelerations in enzymic and intramolecular reactions and the chelate effect. *Proceedings of the National Academy of Sciences of the United States of America* 68, 1678–1683.

77. Villa, J., Strajbl, M., Glennon, T. M., Sham, Y. Y., Chu, Z. T. & Warshel, A. (2000). How important are entropic contributions to enzyme catalysis? *Proceedings of the National Academy of Sciences of the United States of America* 97, 11899–11904.

78. Fothergill, M., Goodman, M. F., Petruska, J. & Warshel, A. (1995). Structure-energy analysis of the role of metal-ions in phosphodiester bond hydrolysis by DNA polymerase I. *Journal of the American Chemical Society* 117, 11619–11627.

79. Florian, J. (2002). Comment on molecular mechanics for chemical reactions. *Journal of Physical Chemistry A* 106, 5046–5047.

80a. Valero, R., Song, L., Gao, J. & Truhlar, D. G. (2009). Perspective on Diabatic Models of Chemical Reactivity as Illustrated by the Gas-Phase S_N2 Reaction of Acetate Ion with 1,2-Dichloroethane. *Journal of Chemical Theory and Computation* 5, 1–22.

80b. Kamerlin, S. C. L., Cao, J., Rosta, E. & Warshel, A. (2009). On Unjustifiably Misrepresenting the EVB Approach While Simultaneously Adopting It. *Journal of Physical Chemistry B* 113, 19095–10915.

81. Kamerlin, S. C. L., Vicatos, S., Dryga, A. & Warshel, A. (2011). Coarse-grained (multiscale) simulations in studies of biophysical and chemical systems. *Annual Review of Physical Chemistry* 62, 41–64.

82. Vorobyov, I., Kim, I., Chu, Z. T. & Warshel, A. (2016). Refining the treatment of membrane proteins by coarse-grained models. *Proteins: Structure, Function and Bioinformatics* 84, 92–117.

83. Warshel, A. (2014). Multiscale modeling of biological functions: from enzymes to molecular machines (Nobel Lecture). *Angewandte Chemie-International Edition* 53, 10020–10031.

84. Mukherjee, S. & Warshel, A. (2011). Electrostatic origin of the mechano-chemical rotary mechanism and the catalytic dwell of F1-ATPase. *Proceedings of the National Academy of Sciences of the United States of America* 108, 20550–20555.

85. Astumian, R. D., Mukherjee, S. & Warshel, A. (2016). The physics and physical chemistry of molecular machines. *ChemPhysChem* 17, 1719–1741.

86. Mukherjee, S. & Warshel, A. (2012). Realistic simulations of the coupling between the protomotive force and the mechanical rotation of the F_0 ATPase. *Proceedings of the National Academy of Sciences of the United States of America* 109, 14876–14881.

87. Alhadeff, R. & Warshel, A. (2017). Reexamining the origin of the directionality of myosin V. *Proceedings of the National Academy of Sciences of the United States of America* 114, 10426–10431.

88. Doyle, D. A., Cabral, J. M., Pfuetzner, R. A., Kuo, A. L., Gulbis, J. M., Cohen, S. L., Chait, B. T. & MacKinnon, R. (1998). The structure of the potassium channel: Molecular basis of K^+ conduction and selectivity. *Science* 280, 69–77.

89a. Roux, B. & MacKinnon, R. (1999). The cavity and pore helices in the KcsA K^+ channel: Electrostatic stabilization of monovalent cations. *Science* 285, 100–102.

89b. Berneche, S. & Roux, B. (2001). Energetics of ion conduction through the K^+ channel. *Nature* 414, 73–77.

90. Åqvist, J. & Luzhkov, V. (2000). Ion permeation mechanism of the potassium channel. *Nature* 404, 881–884.

91. Burykin, A., Schutz, C. N., Villa, J. & Warshel, A. (2002). Simulations of ion current in realistic models of ion channels: The KcsA potassium channel. *Proteins: Structure, Function and Bioinformatics* 47, 265–280.

92. Burykin, A., Kato, M. & Warshel, A. (2003). Exploring the origin of the ion selectivity of the KcsA potassium channel. *Proteins: Structure, Function and Bioinformatics* 52, 412–426.

93. Dryga, A., Chakrabarty, S., Vicatos, S. & Warshel, A. (2012). Realistic simulation of the activation of voltage-gated ion channels. *Proceedings of the National Academy of Sciences of the United States of America* 109, 3335–3340.

94. Kim, I., Chakrabarty, S., Brzezinski, P. & Warshel, A. (2014). Modeling gating charge and voltage changes in response to charge separation in membrane proteins. *Proceedings of the National Academy of Sciences of the United States of America* 111, 11353–11358.

95. Lee, M., Kolev, V. & Warshel, A. (2017). Validating a coarse-grained voltage activation model by comparing its performance to the results of Monte Carlo simulations. *Journal of Physical Chemistry B* 121, 11284–11291.

96. Alhadeff, R., Vorobyov, I., Yoon, H. W. & Warshel, A. (2018). Exploring the free-energy landscape of GPCR activation. *Proceedings of the National Academy of Sciences of the United States of America* 115, 10327–10332.

97. Alhadeff, R. & Warshel, A. (2020). A free-energy landscape for the glucagon-like peptide 1 receptor GLP1R. *Proteins: Structure, Function and Bioinformatics* 88, 127–134.

98. Mondal, D., Kolev, V. & Warshel, A. (2020). Exploring the activation pathway and G(i)-coupling specificity of the mu-opioid receptor. *Proceedings of the National Academy of Sciences of the United States of America* 117, 26218–26225.

99. Hwang, J. K. & Warshel, A. (1987). Semiquantitative calculations of catalytic free-energies in genetically modified enzymes. *Biochemistry* 26, 2669–2673.

100. Roca, M., Vardi-Kilshtain, A. & Warshel, A. (2009). Toward accurate screening in computer-aided enzyme design. *Biochemistry* 48, 3046–3056.

101. Frushicheva, M. P., Cao, J., Chu, Z. T. & Warshel, A. (2010). Exploring challenges in rational enzyme design by simulating the catalysis in artificial kemp eliminase. *Proceedings of the National Academy of Sciences of the United States of America* 107, 16869–16874.

102. Jindal, G., Ramachandran, B., Bora, R. P. & Warshel, A. (2017). Exploring the development of ground-state destabilization and transition-state stabilization in two directed evolution paths of Kemp eliminases. *ACS Catalysis* 7, 3301–3305.

103. Mondal, D., Kolev, V. & Warshel, A. (2020). Combinatorial approach for exploring conformational space and activation barriers in computer-aided enzyme design. *ACS Catalysis* 10, 6002–6012.

104. Marchi, M., Gehlen, J. N., Chandler, D. & Newton, M. (1993). Diabatic surfaces and the pathway for primary electron-transfer in a photosynthetic reaction center. *Journal of the American Chemical Society* 115, 4178–4190.

105. Alden, R. G., Parson, W. W., Chu, Z. T. & Warshel, A. (1995). Calculations of electrostatic energies in photosynthetic reaction centers. *Journal of the American Chemical Society* 117, 12284–12298.

106. Ball, P. (2004). Enzymes — By chance, or by design? *Nature* 431, 396–397.

107. Warshel, A. (1978). Charge stabilization mechanism in the visual and purple membrane pigments. *Proceedings of the National Academy of Sciences of the United States of America* 75, 2558–2562.

108. Schutz, C. N. & Warshel, A. (2001). What are the dielectric "constants" of proteins and how to validate electrostatic models? *Proteins: Structure, Function and Bioinformatics* 44, 400–417.

Acknowledgments

I would like to thank my family for providing me with a great supporting environment.

I am grateful to my co-workers, students and postdocs for their immense contributions to the progress of my works and to my accomplishments.

I greatly appreciate the help of Bill Parson, Yael Warshel, Zhaph and Moshe Rubinstein for insightful comments on the manuscript. I am also grateful for the great editing work of Sharon Ashely. I would also like to acknowledge the influence of my parents that helped in a subliminal way to push me in the direction I took.

About the Author

Dr Arieh Warshel is Distinguished Professor of Chemistry and Biology at the University of Southern California, USA, where he currently holds the Dana and David Dornsife Chair in Chemistry. He was awarded the 2013 Nobel Prize in Chemistry, alongside Martin Karplus and Michael Levitt, for the development of multiscale models for complex chemical systems. He is an elected Member of the USA National Academy of Sciences, a Foreign Member of the Russian Academy of Sciences, Honorary Fellow of the Royal Society of Chemistry, and elected Fellow of the Biophysical Society and the American Association for the Advancement of Science. He has authored over 500 peer-reviewed articles and a book, *Computer Modeling of Chemical Reactions in Enzymes and Solutions* (Wiley, 1991).

Dr Warshel was born in 1940 to a family of founders of the Sde Nahum kibbutz, in what was then Mandatory Palestine (now Israel). He served in the Israeli Signal and Armored Corps, and fought in both the 1967 Six-Day War and the 1973 Yom Kippur War, attaining the final rank of Captain. He attended the Technion in Haifa, where he studied Chemistry based on an off-the-cuff suggestion by a friend. He received his BSc degree, summa cum laude, in 1966 — the same year he married his wife Tamar, with whom he has two daughters. He earned his MSc (1967) and PhD (1969) in Chemical Physics at the Weizmann Institute of Science, working under the Institute's scientific director, Shneior Lifson, and collaborating with Michael Levitt. After his PhD, he did postdoctoral work at Harvard University with Martin Karplus from 1970–1972. He then returned to the Weizmann Institute and also worked at the Laboratory of Molecular Biology in Cambridge, England where he was reunited with Michael Levitt. He joined the Chemistry faculty at USC in 1976, where he has been ever since.

Dr Warshel is known for his work on computational biochemistry and biophysics, in particular for pioneering multiscale simulations of the functions of biological systems, and for developing what is known today as Computational Enzymology. He is responsible for introducing molecular dynamics in biology; developing the QM/MM approach; introducing simulations of enzymatic reactions; pioneering microscopic simulations of electron transfer and proton transfer in solutions and in proteins; pioneering microscopic modeling of electrostatic effects in macromolecules; and introducing simulations of protein folding. He has continued to be active in research, and in April 2017 he opened the Warshel Institute for Computational Biology at the CUHK Shenzhen campus, with the intention of fostering one of the world's most advanced computational biology centers in the Southern Chinese city. As of 2021, he has an h-index of 118.